Recent Aspects of
Nitrogen Metabolism in Plants

First Long Ashton Symposium 1967

Recent Aspects of Nitrogen Metabolism in Plants

Proceedings of a Symposium held at
Long Ashton Research Station
University of Bristol
18–19 April 1967

Edited by E. J. HEWITT *and* C. V. CUTTING

1968

ACADEMIC PRESS · London · New York

ACADEMIC PRESS INC. (LONDON) LTD
Berkeley Square House
Berkeley Square
London, W.1

U.S. Edition published by
ACADEMIC PRESS INC
111 Fifth Avenue
New York, New York 10003

Library of Congress Catalog Card Number: 68-17672

PRINTED IN GREAT BRITAIN BY
J. W. ARROWSMITH LTD
BRISTOL

Participants in the Symposium

ABBOTT, Dr A. J. Long Ashton Research Station, University of Bristol
ANDERSON, Dr J. W. Department of Botany, University College, London
BARBER, Dr D. A. A.R.C. Radiobiological Laboratory, Wantage, Berks
BETTS, Dr G. F. Long Ashton Research Station, University of Bristol
BOND, Professor G. Department of Botany, University of Glasgow
BOULD, Dr C. Long Ashton Research Station, University of Bristol
BOURNE, W. F. Department of Plant Science, University of Newcastle-upon-Tyne
BROWN, Dr D. H. Department of Botany, University of Bristol
BROWN, Professor R., F.R.S. University of Edinburgh
BUSSEY, A. H. Department of Botany, University of Bristol
CHATT, Professor J., F.R.S. A.R.C. Unit of Nitrogen Fixation, The Chemical Laboratory, University of Sussex, Brighton
CLINCH, Professor Phyllis E. M. Department of Botany, University College, Dublin
COCKING, Dr E. C. Department of Botany, University of Nottingham
COX, G. F. Department of Biochemistry, School of Biological Sciences, University of East Anglia, Norwich
COX, Dr Rosalie M. Department of Botany, University of Bristol
CROOKE, Dr W. M. Macaulay Institute for Soil Research, Craigiebuckler, Aberdeen
DAVIES, Professor D. D. School of Biological Sciences, University of East Anglia, Norwich
DEKOCK, Dr P. C. Macaulay Institute for Soil Research, Craigiebuckler, Aberdeen
DODGE, Dr A. D. School of Biological Sciences, Bath University of Technology
DONE, Dr J. Department of Botany, University College, London
DOWDELL, R. School of Biological Sciences, Bath University of Technology
DUNCAN, R. J. S. Clare College, Cambridge
DUNNILL, Dr P. Department of Botany, University College, London
*FOLKES, Dr B. F. Department of Botany, University of Bristol
FOTTRELL, Dr P. F. Department of Biochemistry, University College, Galway, Eire
FOWDEN, Professor L., F.R.S. Department of Botany, University College, London
FREEMAN, Dr G. G. National Vegetable Research Station, Wellesbourne, Warwick

GASSER, Dr J. K. R. Rothamsted Experimental Station, Harpenden, Herts
GOODMAN, Dr P. J. Welsh Plant Breeding Station, University College of Wales, Aberystwyth, Cardiganshire
GRAY, Dr D. O. Department of Botany, University College, London
GREENHAM, Dr D. W. P. East Malling Research Station, Maidstone, Kent
GUNDRY, C. S. Long Ashton Research Station, University of Bristol
HEATH, Professor O. V. S., F.R.S. Department of Horticulture, University of Reading, Berkshire.
HEWITT, Dr E. J. Long Ashton Research Station, University of Bristol
HILL, J. M. Rothamsted Experimental Station, Harpenden, Herts
HILL-COTTINGHAM, Dr D. G. Long Ashton Research Station, University of Bristol
HUCKLESBY, D. P. Long Ashton Research Station, University of Bristol
HUGHES, Dr J. C. A.R.C. Food Research Institute, Earlham Laboratory, Norwich
JEFFREY, Dr D. W. Department of Biology, University of Lancaster
JEWISS, Dr O. R. Grassland Research Institute, Hurley, Berks
JONES, Dr D. PRICE, Jealott's Hill Research Station, Bracknell, Berks
JONES, K. Department of Biology, University of Lancaster
JONES, Dr O. T. G. Department of Biochemistry, University of Bristol
KIRKBY, E. A. Department of Agricultural Chemistry, University of Leeds
LEIGH, Dr G. J. A.R.C. Unit of Nitrogen Fixation, University of Sussex
LINEHAN, D. J. A.R.C. Food Research Institute, Earlham Laboratory, Norwich
LUCKWILL, Dr L. C. Long Ashton Research Station, University of Bristol
MANN, Dr P. J. G. Rothamsted Experimental Station, Harpenden, Herts
†MARKHAM, Dr R., F.R.S. A.R.C. Virus Research Unit, Cambridge
MASTERSON, C. The Agricultural Institute, Johnstown Castle, Wexford, Eire
MIFLIN, Dr B. J. Department of Plant Science, University of Newcastle-upon-Tyne
MONTGOMERY, Phyllis, Department of Biochemistry, University College, Galway, Eire
NEWBOULD, Dr P. A.R.C. Radiobiological Laboratory, Wantage, Berkshire
NOTTON, B. A. Long Ashton Research Station, University of Bristol
NOWAKOWSKI, Dr T. Z. Rothamsted Experimental Station, Harpenden, Herts
OSMOND, Dr B. Department of Botany, University of Cambridge
PATE, Dr J. S. Department of Botany, Queen's University, Belfast
PATTERSON, Dr B. D. Department of Biochemistry, University of East Anglia, Norwich
PAYNE, D. H. Woodstock Agricultural Research Centre, Sittingbourne, Kent

POJNAR, Dr E. Department of Botany, University of Nottingham
POLLARD, Dr A. Long Ashton Research Station, University of Bristol
RENFELL, Gail, Ashburn Hall, Fallowfield, Manchester
ROBB, D. A. School of Molecular Sciences, University of Warwick, Coventry
ROUGHLEY, R. J. Rothamsted Experimental Station, Harpenden, Herts
SIMS, Dr A. P. Department of Botany, University of Bristol
SLAUGHTER, C. Department of Biochemistry, University of East Anglia, Norwich
SMITH, I. K. Department of Botany, University College, London
SMITH, Dr T. A. A.R.C. Systemic Fungicides Unit, Wye College, Ashford, Kent
SPRATT, E. D. Rothamsted Experimental Station, Harpenden, Herts
STEWART, G. R. Department of Botany, University of Bristol
‡STREET, Professor H. E. Department of Botany, University College of Swansea
THAINE, R. Grassland Research Institute, Hurley, Berks
TREWAVAS, Dr A. Department of Biochemistry, University of East Anglia, Norwich
TRUTER, Dr. Mary, A.R.C. Unit of Structural Chemistry, University College, London
TUNNEY, H. The Agricultural Institute, Johnstown Castle, Wexford, Eire
WAID, Dr J. G. Department of Soil Science, University of Reading
WALDEGRAVE, The Rt. Hon. the Earl, Chewton House, Chewton Mendip, Somerset
WALKER, Christine, Department of Botany, University College of Swansea
WHATLEY, Professor F. R. Department of Botany, King's College, London
WHITING, Dr G. C. Long Ashton Research Station, University of Bristol
WILLIAMS, A. H. Long Ashton Research Station, University of Bristol
WILLIS, Dr A. Department of Botany, University of Bristol
WINSOR, Dr G. W. Glasshouse Crops Research Institute, Littlehampton, Sussex
YATES, Dr M. G. A.R.C. Unit of Nitrogen Fixation, University of Sussex
YEMM, Professor E. W. Department of Botany, University of Bristol
YOUNG, M. Department of Botany, University of Bristol

Director
Professor H. G. H. KEARNS, O.B.E.

Scientific Liaison Officer
Dr C. V. CUTTING

*Now Professor at the School of Biological Sciences, University of East Anglia.

†Now Professor of Cell Biology at the University of East Anglia, and Director of the John Innes Institute, Norwich.

‡Now Professor of Cell Biology, University of Leicester.

Preface

This book contains the papers presented at the first of a series of biennial symposia planned by the staff of Long Ashton Research Station, University of Bristol. It provides a well-documented account of the present intensive research into nitrogen assimilation in plants.

The contributions of Lord Waldegrave and Professor Kearns explain adequately the origin of this symposium, and its significance in the wider field of agricultural practice. In addition, we wish to make our own acknowledgements to the contributors for complying so fully with our request for accounts of new research findings set against a background of appropriate reviews. This seemed desirable as it is now several years since this rapidly-developing subject was surveyed in Britain, at the Reading Conference of the Society for Experimental Biology in 1959.

The symposium was divided into sections dealing with biochemical and chemical aspects of nitrogen fixation and metabolism, and physiological and external factors affecting the utilization of the element.

We hope the wide scope of the book will satisfy the needs of the specialist worker in this subject, and provide him with an up-to-date source guide, and that those engaged in other aspects of biochemistry will find lucid accounts of advances which may be relevant to their own studies. Scientists concerned with the wider sphere of agricultural and horticultural practice should discover much that is enlightening and of economic significance.

We appreciate very much the co-operation of contributors in their early preparation of papers which allowed adequate pre-circulation and permitted more valuable discussion. We are grateful to the Chairmen who presided, and for their able leadership of the discourses. The symposium owes a debt to some of its younger members for the careful recording and transcription of the discussions. We would like to thank the many typists, both our own at Long Ashton and those of the contributors, for their help.

Above all, we value the assistance of the Editorial and Production staff of Academic Press, who have dealt so well with our varied material and expedited publication.

We also gratefully acknowledge the permission granted by the journals concerned for reproduction of several diagrams and tables, as indicated in the captions.

30 November, 1967

E.J.H.
C. V. C.

Foreword

Address by

The Rt. Hon. the Earl Waldegrave, T.D., D.L.

Chairman of the Agricultural Committee, University of Bristol

I am very glad to be here but am very embarrassed. Today is the age of the professional, and the amateur, whether gifted or otherwise, is said to have had his day. Today is the age of common man, *homo vulgaris* perhaps rather than *homo sapiens*! Today is certainly not the age of Disraeli's Aristocratic Principle. And of course it is a young man's world now—brown beards, not grey heads are in fashion. Yet here am I, an elderly non-scientific, hereditary peer, a square to the power of n. But perhaps it is all right as no less a person than Sir Solly Zuckerman has said that "we live in an age of paradox".

This symposium is of first class importance and interest. I am credibly informed that the papers are of a very high scientific level and we are indeed fortunate that we have attracted such very distinguished speakers to read papers for us. The first thing I must do is to thank Professor Kearns, and all the people who work with him at Long Ashton, for all he has done in bringing this symposium about. A great deal of work has to be done behind the scenes before an occasion like this, and the proof that this work has been well done is the smoothness of the organization when the event takes place. We are extremely grateful to you, the contributors, for coming here, and I know that you will wish me on your behalf to thank the organizers for the excellence of the arrangements they have made. You will be glad to know that the Proceedings are to published so that others can benefit from your deliberations.

I think it would be appropriate also for me to remind you that this symposium has been made possible by a grant from the National Fruit and Cider Institute which was the parent of this great Research Station. I hope that from time to time it may be possible to arrange similar symposia in the future with grants from the Institute, though I should add that funds available for this purpose are regrettably limited.

Your programme has been arranged so that you are able to develop the subject comprehensively from the basic consideration of inorganic nitrogen fixation and metabolism to the rather more applied aspects of the subject which you have in Section 3. The Director tells me in fact that the intention of the symposium, which no doubt by now you will have recognized, is to relate the fundamental chemistry of nitrogen to the biochemical mechanisms by which it is transformed and utilized, and to relate these mechanisms to the physiological conditions which influence their activity. It is always easier for a non-scientific, lay tax-payer to be sincere in his praise for scientific work when he can see obvious material advantage likely to accrue. The meanest classical scholar can dimly comprehend that water affects plant growth and that plants produce food. Non-scientists a little further up the scale of intelligence can comprehend that nitrogen is an essential element in plant and animal metabolism. Indeed the study of nitrogen requirements in plants is one of the oldest aspects of plant nutrition, and the significance of nitre, as it was then called, was recognized in the seventeenth century. Now, some 300 years later, the study of the factors affecting the activity and production of the enzymes involved in the reduction of nitrates and nitrites to ammonia is a major part of the work of this Research Station at Long Ashton.

When we think about the vast needs for food which may indeed overwhelm us if the population continues to increase at its present rate, we cannot help but appreciate the absolute necessity of continuing high level, fundamental research on something so essential to life as nitrogen, and having recognized the necessity of the research, we shall be more ready, indeed almost happy, to pay for it. Enormous quantities of nitrogen supplies are already being manufactured by expensive and laborious processes, and we have not scratched the surface of our future needs. All the time, nature is converting atmospheric nitrogen by simple fixation without the aid of man-made factories.

So, ladies and gentlemen, if you can unravel these natural processes—rather, *when* you have unravelled these natural processes, you will deserve a very great deal more than you will probably be given!

Introduction

Professor H. G. H. Kearns, O.B.E., Ph.D., D.Sc., Dip. Agric.

For those who are visiting Long Ashton for the first time, I should explain that although we are a Department of the University, we have no teaching responsibilities to undergraduates. Our previous history is a rather interesting one. We are the oldest fruit research station and are nearly 65 years old. We were started, in fact, by private enterprise. There was an interesting character, R. Neville Grenville, who lived at Butleigh Court, near Glastonbury in mid-Somerset, who found that sometimes cider and perry was superb but often was of poor quality. At one time there were nearly 150,000 acres of cider apples in the West Country because cider provided the essential lubricant to the manual labour of those days. There was, as you know, in the early nineteenth century a great wheat belt in South Somerset, all of which was hand harvested, so farm labourers had a ration of one gallon a day and it is not surprising that many of them died young when you bear in mind that cider was produced by uncontrolled fermentations. Grenville decided that it would be a very good thing to put some research into cider making and spent his own money on this problem at Butleigh Court. He very soon found, however, that as he was not a scientist he could not produce a consistently good brew, and so he persuaded a number of his friends to subscribe to a fund to establish a cider research institute, and that is how we came to Long Ashton in 1903.

The Board of Agriculture of that time, who were suspicious of new-fangled research, thought that they should be at least on the edge of the band wagon and gave a small grant. Very soon we were firmly established under Barker who became Director. His research went from strength to strength; people became interested and it was soon clear that we had to give attention to other topics, such as cider pomology and pest and disease control, and in 1912 we had grown to sufficient stature and became the Department of Agriculture and Horticulture of the University. From that time we steadily enlarged our sphere of interests and now have a staff of some 250, with Cider and Fruit Juices as our largest section.

You can well imagine, as the Station developed, we became interested in more disciplines. For example, the late Professor Wallace, my predecessor, was an authority on plant nutrition and he gave the department an international fame by his research on fruit crops. During the course of this work it became clear that we had to embark on fundamental research and one of the important developments, supported by a Unit of the Agricultural Research Council, was the investigation of mineral deficiencies and micronutrients. A research group including Hewitt, Nicholas and several others became particularly interested in nitrogen metabolism. I think these gentlemen, who in this later phase spent almost their whole time in the laboratory on these complex problems, might feel lost in our experimental plantations. They are not particularly concerned in the practical techniques of crop culture, but they are very much interested in what happens to nitrogen. As you all know, nitrogen is one of the great kingpins for both quantity and quality in crop production, particularly so in the case of the apple and the black currant, of which you will probably hear further later on.

Gentlemen, I am not going to waste a lot of your time because I know you have a heavy programme of papers. I would just like to say that this symposium was conceived by Hewitt and his colleagues, and we felt that although we ourselves could only make a crumb of contribution to this very important field, we dare try to arrange a symposium provided eminent authorities were able to help us and we are very grateful to you all for the contributions you are about to make. This symposium would not have been possible without your enthusiastic support.

We are also greatly indebted to the University Agricultural Committee, under the Chairmanship of Lord Waldegrave, who have made a generous grant by using the funds of the National Fruit and Cider Institute to finance, in part, this symposium. It is now my pleasure to welcome you to Long Ashton and to invite you to begin your meeting.

Contents

SECTION 3

Interaction of Nitrogen Metabolism with External Factors

Chairman: Professor O. V. S. Heath, F.R.S.
University of Reading

Section 1

Transformations in Inorganic Nitrogen Metabolism and Fixation

Section 1a

The Inactivity and Activation of Nitrogen

J. CHATT AND G. J. LEIGH

Agricultural Research Council,
Unit of Nitrogen Fixation,
The University of Sussex, Brighton, England

THE INACTIVITY OF NITROGEN

The nitrogen molecule is uniquely inert. All other diatomic molecules are reactive entities, e.g. O_2, F_2, NO and CO. The nitrogen molecule, written with the valence bond structure $N \equiv N$, is closely similar to acetylene, $HC \equiv CH$, but whereas acetylene is highly reactive and even explodes on compression to two atmospheres pressure, nitrogen resists all except the most reactive of reagents (e.g. Li metal) at ordinary temperature. Nevertheless, *a priori* the nitrogen molecule appears to have plenty of places for attack. The triple bond in acetylene is highly reactive, but the triple bond in nitrogen, seemingly very similar, confers no reactivity. Carbon monoxide, iso-electronic with nitrogen, owes its reactivity to the non-bonding electron pair on the carbon atom. Nitrogen, too, appears to have lone pairs, one on each nitrogen atom, yet it is inactive. We must, therefore, enquire why nitrogen gas is so inert, and when we know the reason, how to activate it.

NITROGEN AND ITS ANALOGUES

Nitrogen is a closely knit molecule with a very high energy of dissociation (224·5 kcals). In the case of nitrogen, which is diatomic, the dissociation energy is equal to the bond energy, and Table I lists some relevant bond energies for comparison.

TABLE I

Bond energies (kcals).

	C–C	C–N	N–N
single	82·6	72·8	38·4
double	145·8	147·0	97·6
triple	199·6	212·6	224·5

The strength of the N–N triple bond is reflected also in other physical properties of the nitrogen molecule. Some of these are listed in Table II

TABLE II

Some physical constants of nitrogen and its analogues.

	$(HC)_2$	N_2	O_2	CO	NO
Bond length (Å)	1·208	1·098	1·207	1·128	1·150
Ionization potential (eV)	11·4	15·6	12·3	14·0	9·25
Dissociation energy (kcals)	199·6*	224·5	117·8	256·2	150·0
Stretching frequency (cm^{-1})	1974	2331	1555	2143	1876

* Bond energy.

together with similar properties of related molecules. It will be noted that nitrogen has the shortest bond length, the highest ionization potential, and the highest stretching frequency; its dissociation energy is exceeded only by that of carbon monoxide. The only stable similar diatomic species with a shorter bond length is NO^+ (1·062 Å), and here there is a positive charge to draw in the electrons. The nearest analogue to nitrogen is carbon monoxide, which has an even higher dissociation energy. Nevertheless, it is easier to oxidize and more reactive because of the non-bonding electron pair in an essentially *sp*-hybridized orbital. This lone pair on carbon accounts for its lower ionization potential. Acetylene has a low ionization potential because its electrons in the π-bonding orbitals are relatively weakly held. Nitric oxide and oxygen have low ionization potentials because they have electrons in antibonding π-orbitals. These account for the high reactivities and relatively low bond strengths of these molecules. The stretching frequencies observed in the Raman spectra are also a measure of bond strength. Acetylene appears anomalous and, indeed, substituted acetylenes, RC≡CR (R = alkyl groups) which are also highly reactive, have higher stretching frequencies (2260–2190 cm^{-1}). The electrons in the nitrogen molecule are exceptionally tightly held, the ionization potential (15·58 eV) is almost the same as that of argon (15·76 eV) which is well known to form no chemical compounds.

These unique properties of the nitrogen molecule must reflect a unique structure, so we must now enquire how the bonding in a nitrogen molecule is unique and, especially, how it differs from that of acetylene, of carbon monoxide and of oxygen. Most chemists are familiar with the bonding scheme of acetylene. Here we have a linear σ-bonded H–C–C–H skeleton with six electrons in low energy molecular orbitals formed by

2*sp*-hybridized orbitals on each carbon atom, and the 1*s*-orbitals on the hydrogen atoms. In addition, there are two bonding π-molecular orbitals, formed by mixing of the remaining two *p*-orbitals on each carbon atom with the corresponding parallel *p*-orbitals on the other carbon atom (as the $1\pi_u$-orbital of N_2 in Fig. 4). These contain four electrons and are on a much higher energy level than the σ-system. They are responsible for the reactivity of acetylene. In acetylene all bonding orbitals are filled and there are no electrons in high energy antibonding orbitals.

BONDING IN THE NITROGEN MOLECULE

We can understand how acetylene differs from nitrogen if we consider the changes in atomic energy level which occur on passing along the elements in the first row of the Periodic Table. Two important energy changes take place: (a) the atomic orbital energy falls rapidly; (b) the energy separation of the 2*s*- and 2*p*-atomic orbitals rapidly increases. The exact energies of the atomic orbitals are not certain but, however they are calculated, the above two changes are found. Figures 1 (Jaffé, 1956) and 2 (values from

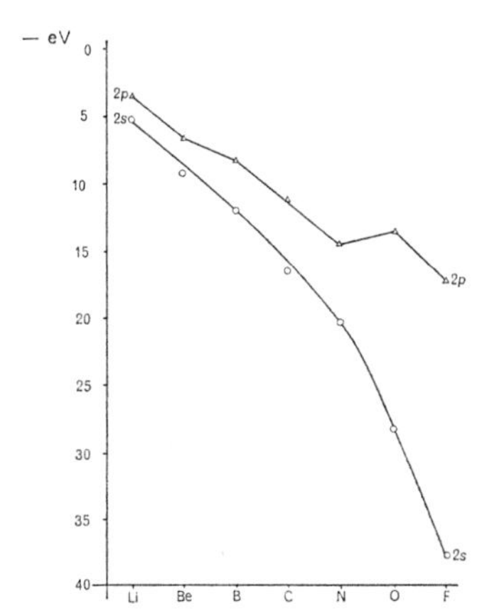

FIG. 1. Approximate energy levels of *s*- and *p*-orbitals.

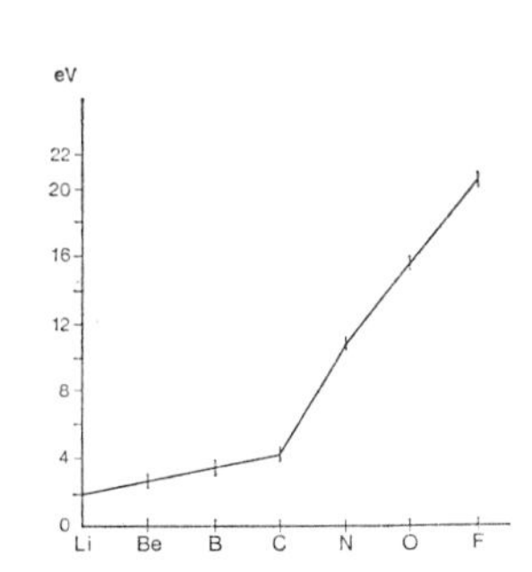

FIG. 2. Approximate 2*s*–2*p* energy separation.

Moore, 1949), which are not quite self-consistent, show how the two factors (a) and (b) vary with atomic number across the first short period. These changes in the energy levels have two important consequences: (a) the electrons are held more strongly with increasing atomic number (to fluorine); (b) there is less *sp*-hybridization because the mixing of orbitals occurs best between orbitals of equal energy.

Very little *sp*-hybridization occurs on the nitrogen, oxygen or fluorine atoms. Thus the nitrogen molecule differs from acetylene; it does not have a linear *sp*-hybridized σ-bonded skeleton, flanked by two high energy π-molecular orbitals containing electrons. In the nitrogen molecule the bonds are formed by the overlap of almost pure *s*- and pure *p*-orbitals as shown in the correlation diagram (Fig. 3), which also shows the measured

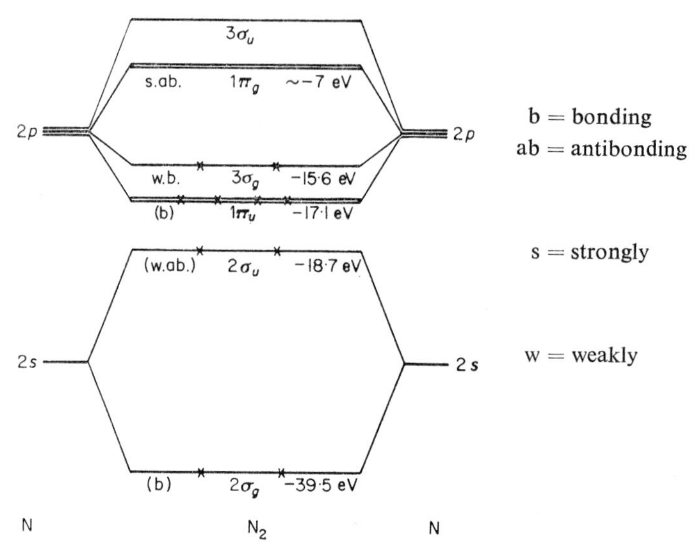

FIG. 3. Energy levels in the nitrogen molecule.

energy levels of the molecular orbitals relative to the energy of a free electron which is taken as zero (see Mulliken, 1958). The 1*s*-orbitals, which have very low energies and contribute nothing to the bonding are omitted. The shapes of the molecular orbitals are shown schematically in Fig. 4.

The nitrogen molecule is distinguished from all its analogues in that the highest energy orbital ($3\sigma_g$) of ionization potential 15·6 eV is a σ-bonding orbital. It is only weakly bonding. The main bonding energy in the nitrogen molecule arises from the π-bonding orbitals ($1\pi_u$). The so-called lone pairs of electrons on the nitrogen atoms are found in very low energy orbitals derived from the *s*-atomic orbitals ($2\sigma_g$ and $2\sigma_u$). These electrons are very tightly held. Each nitrogen atom contributes five electrons to the molecular orbital system shown in Fig. 3, i.e. ten electrons altogether, just sufficient to fill all five orbitals up to and including the $3\sigma_g$. Thus all the bonding orbitals are filled and the vacant orbital of lowest energy is an antibonding π-molecular orbital ($1\pi_g$ at -7 eV).

The nitrogen molecule has all its electrons in low energy orbitals (highest energy $-15{\cdot}6$ eV). Its analogues all have electrons in higher energy orbitals, e.g. acetylene in the π-bonding orbitals ($-11{\cdot}4$ eV) and carbon monoxide in the non-bonding orbital on carbon ($-14{\cdot}0$ eV). In NO, O_2

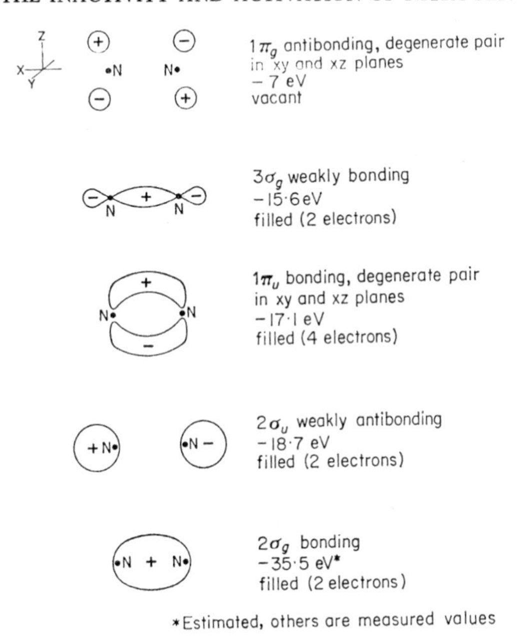

FIG. 4. Schematic representation of important molecular orbitals in the nitrogen molecule.

and F_2 the number of electrons in the molecule exceeds that in the nitrogen molecule by 1, 2 and 4 respectively, and these additional electrons must be accommodated in the two degenerate antibonding orbitals, $1\pi_g$, so weakening the π-bonds, causing the interatomic distance to increase, and reversing the relative energies of the $1\pi_u$ and $3\sigma_g$ molecular orbitals; thus in oxygen, as in acetylene, the bonding π-molecular orbitals are found at a higher energy level than the bonding σ-molecular orbitals. The inactivity of nitrogen resides in its unique electronic structure.

THE ACTIVATION OF NITROGEN

OXIDATION AND REDUCTION

Nitrogen is unique amongst its well-known analogues in having as its highest filled molecular orbital a σ-orbital buried in the centre of the molecule, and in having very low energy electrons in its π-molecular orbitals. How can this structure be attacked? Can nitrogen be oxidized, i.e. can it behave as an electron donor? If it can, the electron must come from the filled orbital of highest energy ($3\sigma_g$ at $-15{\cdot}6$ eV). Thus nitrogen is about as difficult to oxidize at room temperature as is argon. At very high temperatures in an excited state nitrogen can be oxidized, and the ion N_2^+ has been observed in discharge tubes, but there is little hope of oxidizing nitrogen at room temperature.

Can nitrogen be reduced, i.e. can it behave as an electron acceptor? Any electron accepted by the nitrogen molecule must go into the antibonding $1\pi_g$ orbital at -7 eV. Strong reducing agents, e.g. the alkali metals which have ionization potentials (eV) Li, 5·390; Na, 5·130; K, 4·134; and Cs, 3·893 should be able to achieve this, and the N_2^- ion so formed with an electron in an antibonding π-orbital should be more weakly bonded than N_2 (compare NO, which is iso-electronic with N_2^-). It is well known that bright lithium wire under nitrogen becomes coated with a dark purple brown nitride. Since the normal LiN_3 is a red solid, we have recently re-investigated this so-called nitride and find that it reacts with water to form ammonia and hydrazine. Obviously it contains some N_2^- or N_2^{2-} which, on hydrolysis, should give some hydrazine by disproportionation of the unstable N_2H or N_2H_2.

It is obvious that neither pure oxidation nor pure reduction can be involved in the biological fixation of nitrogen. No oxidizing agent is sufficiently strong to oxidize nitrogen at room temperature, not even fluorine. No reducing agent strong enough to reduce nitrogen could exist in an aqueous environment, because water would be preferentially reduced to hydrogen. It seems that only a concerted electron acceptor and electron donor mechanism can be postulated for the attack of nitrogen by mild reagents, and several recent, accidental discoveries have pointed the way.

NITROGEN AS A CONCERTED ELECTRON DONOR AND ACCEPTOR

As an electron donor, nitrogen must supply electrons from its $3\sigma_g$ orbital. This is a bonding σ-orbital of very low energy ($-15·6$ eV); the main electron density in this orbital lies between the nitrogen nuclei, but some of it lies outside the two nuclei along the axis of the molecule. Thus, if the nitrogen molecule is to be attacked by an acceptor ion or group, it should be along the axis of the molecule (x-axis in Fig. 4). Metal ions (M) are good electron acceptors, and the process involving an ion could be represented as in Fig. 5.

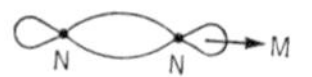

FIG. 5. Nitrogen as an electron donor to a metal ion (M).

When nitrogen is an acceptor, electrons must enter the $1\pi_g$ orbital at -7 eV. To provide the electrons the reagent must have a donor orbital of somewhat similar energy to the $1\pi_g$ orbital of nitrogen and of suitable symmetry. Both *p*- and *d*-atomic orbitals have suitable symmetry and a transition metal ion with electrons on a suitable energy level in its *d*-orbitals could act as an electron donor to the nitrogen. Either of the arrangements

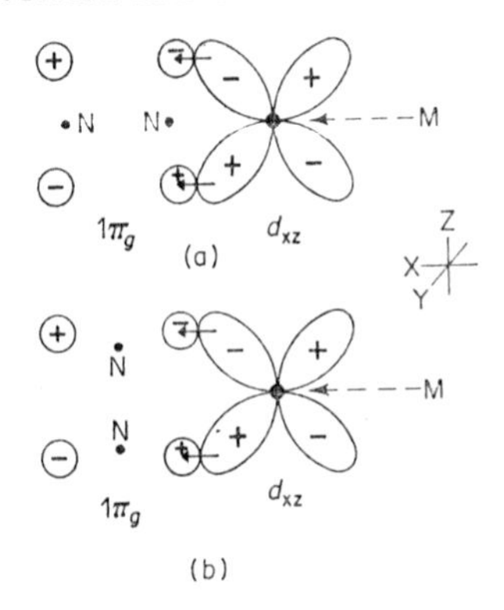

FIG. 6. Nitrogen as an electron acceptor from a transition metal ion (M).

(Fig. 6(a) or 6(b)) which show schematically the shapes of the orbitals involved would be possible. However, there is an identical set of orbitals in the xy plane of Fig. 6(a) but not of Fig. 6(b). This must favour the end-on arrangement of Fig. 6(a). Moreover, if the nitrogen is to act as a donor to the metal, as in Fig. 5, and in concerted action as an acceptor of electrons from it, then the arrangement in Fig. 6(a) is the only one possible. Thus we obtain a bonding system having a nitrogen molecule attached end-on to a transition metal ion. The metal ion must have electrons in non-bonding *d*-orbitals (e.g. d_{xz}) to donate to the nitrogen, which means that it must be in a low oxidation state. The bonding between the metal and the nitrogen molecule will be helped by a synergic effect, whereby electron drift from the *d*-orbitals of the metal ion into the antibonding molecular orbitals of the nitrogen will enhance the reverse flow of electrons from the σ-orbital of the nitrogen into the acceptor orbital of the metal. The metal is, in fact, removing electron density from the bonding σ-orbital and putting electron density into the antibonding π-orbital. Both of these operations will weaken the nitrogen molecule, and render it more susceptible to attack. In fact, the nitrogen molecule will be in a similar condition to the N_2 group in the well-known compounds which are shown in Fig. 7.

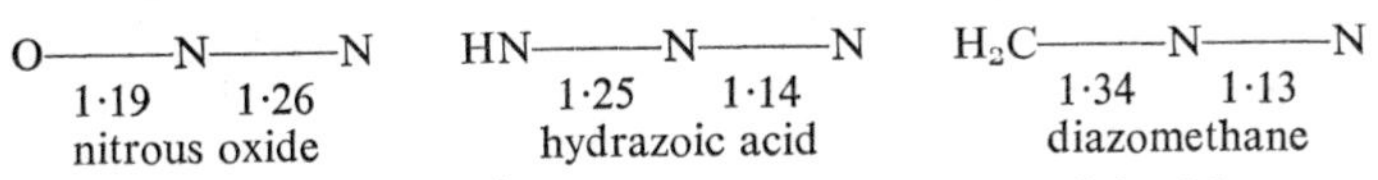

Fig. 7. Bond distances (Å) in common molecules containing N_2 groups.

In recent months four groups of workers have found, by accident, compounds containing a nitrogen molecule bound to a transition metal, apparently in the manner suggested. Allen and Senoff (1965) first reported a series of salts of the ruthenium-containing anion $[Ru(NH_3)_5(N_2)]^{2+}$. The anion is formed by the reaction of hydrazine on ruthenium trichloride in aqueous solution and has a strong band in the infrared spectrum at

2112–2167 cm^{-1}, depending on the nature of the anion. This strong band is assigned to the N–N stretching frequency which is considerably lower than that observed in the N_2 molecule at 2331 cm^{-1}. It shows that the N–N bond is still a triple bond, but rather weakened by attachment of the nitrogen to the metal. Bottomley and Nyburg (1966) have examined crystals of the dichloride of this anion by X-rays, but, unfortunately, the structure is disordered. Nevertheless, it was shown that the Ru–N–N atoms lie in a straight line, and that the distance between the nitrogen atoms of the N_2 entity is about 1·12 Å, just slightly shorter than in diazomethane. The nitrogen in the ruthenium compound can be reduced to ammonia by sodium borohydride. Thus, if the ruthenium compound could be made directly from nitrogen gas in the presence of water containing sodium borohydride, we would have a mild method for reduction of nitrogen and a model for nitrogenase.

A second nitrogen complex, $[IrCl(N_2)(PPh_3)_2]$ which, this time, is non-ionic, was reported by Collman and Kang (1966). It was obtained by the following reaction:

$$[IrClCO(PPh_3)_2] + PhCON_3 \xrightarrow[0°C]{\text{moist } CHCl_3} [IrCl(N_2)(PPh_3)_2] + PhCONCO$$

This compound has an infrared band at 2095 cm^{-1} and the nitrogen can be displaced by other donor molecules.

The third metal complex, and the first to be prepared directly from gaseous nitrogen, was reported in January 1967 by Yamamoto *et al.* (1967). It has been formulated $[Co(N_2)(PPh_3)_3]$, and has a strong band in the infrared spectrum at 2088 cm^{-1}. This was obtained by the reaction of diethylaluminium ethoxide, $AlEt_2OEt$, which probably serves primarily as a reducing agent, on a suspension of cobalt tris-acetylacetonate in a solution of triphenylphosphine in ether. It is a yellow-orange solid, moderately stable in air and loses its nitrogen as N_2 at 80°C.

We have repeated all the above preparations.

Finally, Sacco and Rossi (1967) reported that a cobalt hydride $[CoH_3L_3]$ (L = PPh_3 or $PEtPh_2$) made by the reduction of $[CoL_2X_2]$ (X = Cl, Br or I) with sodium borohydride in presence of the free ligand L in ethanol, absorbed nitrogen to give the complex $[CoH(N_2)L_3]$. It has a strong absorption band at 2080 to 2084 cm^{-1} in the infrared spectrum and, on heating to 100–150°C, liberates one molecule of nitrogen and half a molecule of hydrogen for each molecule of complex. The nitrogen can also be displaced by other ligands of strong coordinating power. It is interesting that this cobalt compound was formed in a protic solvent. It gives us a good working model for nitrogenase. If it could be reduced to ammonia as can the ruthenium complex, then we have a possible catalytic system as shown in Fig. 8, and this could well serve as a model for the natural system. The cobalt complex $[CoH_3L_3]$ is the nitrogenase; it takes in nitrogen gas, rendering the nitrogen susceptible to reduction by a hydrogenase, so producing ammonia and regenerating the original cobalt hydride complex or nitrogenase, thus completing the cycle.

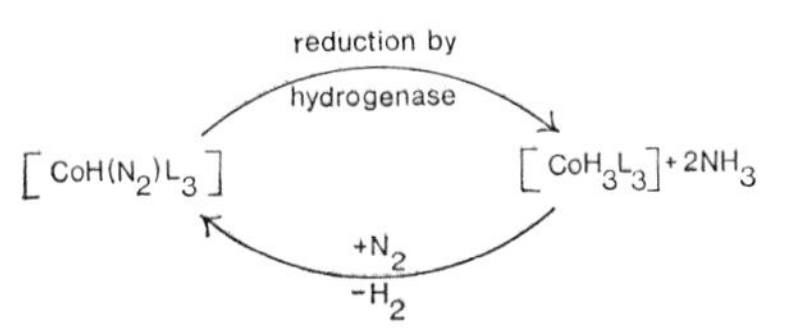

FIG. 8. $[CoH_3L_3]$ as a model for nitrogenase.

Nature does not use complex compounds of the types listed above, but the metal atom in the natural nitrogenase may be in a very similar condition to the metal atom in any of these complexes. It seems unlikely that molybdenum(V), for long considered to be the catalytic metal ion, could form nitrogen complexes of the above types, because molybdenum(V) has only one electron in its non-bonding *d*-orbitals, whereas all the above compounds have all their non-bonding *d*-orbitals filled. This requires six to eight electrons, according to the complex, and essentially it is these electrons which initiate the reduction of the nitrogen molecule.

Thus we can see clearly in principle how nitrogen can be reduced under very mild conditions. We need an activation of the nitrogen molecule by some system with atomic energy levels of suitable energy and symmetry to attack the end of the molecule by withdrawing electrons from its bonding $3\sigma_g$ orbital and putting electrons into its $1\pi_g$ orbitals, this to be followed by reduction to ammonia of the polarized N_2 group so formed. Transition metal ions have electronic energy levels of suitable symmetry but not usually of suitable energy. Evidently the energies are somewhat critical. Nevertheless, by suitable combinations of ligands and metal, the energies can be adjusted to hold nitrogen already attached, as in $[Ru(NH_3)_5(N_2)]^{2+}$ and $[IrCl(N_2)(PPh_3)_2]$ or, as in $[CoH_3(PPh_3)_3]$, to take on nitrogen from the gaseous phase, even in alcohol, which we would normally regard as a better ligand than nitrogen. Nature has obviously solved this problem of adjusting the metal energy levels in such a manner that nitrogen is a preferred ligand, even in competition with water.

If this view is correct, a transition metal ion from Groups IV, V, VI or VII, to have sufficient electrons in its *d*-orbitals, would have to be in so low an oxidation state that it is unlikely to be involved in the natural activation of nitrogen. Those of Group VIII can reach a suitable electronic configuration in their normal oxidation states, e.g. Fe(II), Co(II), Ni(II) and any lower oxidation states. It seems most likely that one of these is the metal used in nature.

After the nitrogen molecule has become attached to the metal, reduction might proceed by electron transfer to the metal complex. These electrons could be fed into the antibonding $1\pi_g$ orbitals of the nitrogen molecule which would then pick up protons from the solution. Alternatively, the reduction could occur by direct attack on the complexed N_2 entity by a hydride complex with transfer of hydride ion. This could well be a hydridic complex of molybdenum, but at this stage of our knowledge such ideas are purely speculative. Suffice it to say that the discovery of nitrogen complexes has at last opened up the study of nitrogen fixation under mild conditions to the chemist. For the first time he can set out with as clearly marked a path as his microbiological and biochemical colleagues.

REFERENCES

Allen, A. D. and Senoff, C. V. (1965). *Chem. Commun.* 621.
Bottomley, F. and Nyburg, S. C. (1966). *Chem. Commun.* 897.
Collman, J. P. and Kang, J. W. (1966). *J. Amer. chem. Soc.* **88**, 3459.
Jaffé, H. H. (1956). *J. chem. Ed.* **33**, 25.
Moore, C. E. (1949). *Atomic Energy Levels*, **1**, Circular 467. Publ. U.S. Dept. of Commerce, National Bureau of Standards.
Mulliken, R. S. (1958). *Can. J. Chem.* **36**, 10.
Sacco, A. and Rossi, M. (1967). *Chem. Commun.* 316.
Yamamoto, A., Kitazume, S., Pu, L. S. and Ikeda, S. (1967). *Chem. Commun.* 79.

Section 1b

Some Biological Aspects of Nitrogen Fixation

G. BOND

Department of Botany
University of Glasgow, Scotland

In the realm of nitrogen fixation the greatest interest at present is in the current efforts to unravel the chemistry of the process, with all the economic benefits that may flow from that, but more broadly based studies seeking to discover new examples of nitrogen-fixing plants or organisms, and to add to knowledge of the physiology and biology of confirmed fixers still have, I trust, some value and interest.

In view of the time factor and of the inevitable limitations in the breadth of one's experience and reading, in this Section I propose to consider mainly fixation by symbiotic systems. It is agreed that in terms of intensity of fixation this is more important than fixation by free-living organisms, though it has to be remembered that the latter are more widespread in their occurrence. I would also explain that my own interest is chiefly in sources of nitrogen in natural rather than agricultural communities of plants.

TABLE I

Symbioses which are certainly or possibly nitrogen-fixing.

Nodulated legumes
Non-legumes with root nodules of *Alnus* type
Cycad root nodules
Azolla–Anabaena association
Liverwort–blue-green alga association
Lichens involving blue-green algae
Leaf nodules
Phyllosphere association
Mycorrhiza:
Podocarp nodules
Other types

The main examples of symbioses with which fixation is known to be, or is possibly, associated are listed in Table I. Though the term "symbiotic fixation" is not actually used, the implication is that it would be extended to all these examples. Stewart (1966) in his recent monograph prefers to restrict its use to instances where the fixation is shown only by the symbiosis, and by neither symbiont when growing alone. Since it is really only the legumes that are definitely known to come into this category it is possible to think that at present sub-division along these lines is unnecessary.

It is proposed to review recent work on the symbioses listed in Table I.

NODULATED LEGUMES

Further electron microscope studies (Burris, 1966) have increased the range of legume nodules examined in this way, and additional information on infection processes and on the fine structure of the infected cells has been provided. The site of fixation within the infected cell, studied by means of the separation of physical fractions of nodules previously exposed to labelled nitrogen, remains somewhat uncertain (Burris, *loc. cit.*). Bergersen (1965) reported that a brei prepared under anaerobic conditions from soya bean nodules still showed clear ability to fix nitrogen when tested isotopically, provided that a low level of oxygen was now furnished. A novel solution to the problem (see above) of why fixation is only shown when legume and rhizobia are associated has been proposed by Turchin *et al.* (1963), in a paper which has received little attention from reviewers. By extraction of frozen, powdered lupin nodules with phosphate buffer, or with ethanol, enzyme preparations active in nitrogen fixation were obtained. Thus when, after addition of pyruvate and ATP, the preparations were exposed for 15–30 minutes to nitrogen with only 4·3 atom % ^{15}N, subsequent analysis of the nitrogen of the ammonium and amide fraction now associated with the preparation showed enrichment as high as 1 atom %. The residual nitrogen of the sample showed no enrichment. The authors then argue that the enzymes of fixation cannot arise *de novo* in the nodules, and that they must be present in legume tissues and/or rhizobial cells constantly but that some factor inhibits their action except in the symbiotic condition. In fact they succeeded in making preparations by the above methods from lupin roots and leaves and from lupin rhizobia in pure culture which showed active fixation. Then, bearing in mind that fixation is associated with a wide range of plants when in symbiosis (Table I), they conceived that perhaps all plants contained fixation enzymes. Active preparations were obtained from leaves of cereals, birch and rose. Schanderl advanced similar ideas some years ago but this is the first time that ^{15}N evidence has been provided in support.

The ingenious experiments of Pate and his associates (Pate and Wallace, 1964; Pate *et al.*, 1965) have added much to knowledge of translocatory arrangements in legumes, while after long neglect the difficult question of the cost to the legume, in terms of photosynthates consumed, of setting up and maintaining the nitrogen-fixing mechanism has been taken up by Gibson (1966). This same question arises in other symbioses where a heterotrophic endophyte is maintained by a photosynthetic partner. His method is that used by some earlier workers, but with improvements in planning and interpretation. Two sets of plants are grown, one nodulated and dependent on nodule nitrogen, the other set without nodules and fed with combined nitrogen at a rate matching that of fixation in the first set. Assuming that photosynthesis proceeds at the same rate in both sets, then the extent to which the dry weight of the nodulated plants falls short of that of the other set after a suitable period is a measure of carbohydrate consumption in the nodules. The clover plants were grown under optimal conditions in growth rooms. During earlier stages, when nodule development was very active, values of 3–10 mg carbohydrate consumed/mg nitrogen fixed were obtained, falling later to 0·8 mg, which is much lower than obtained by previous workers. Bond (1941) relied chiefly on the measurement of the respiration of nodules still attached to the plant, and concluded that some 19 mg carbohydrate was used for each mg nitrogen fixed. In the soya bean plants used by him, nodule dry weight accounted for 8·5% of whole plant weight, but in Gibson's plants for only approximately 2·5%; moreover the clover nodules/unit weight were much more active in fixation. Obviously the results obtained in such tests will vary between different legumes and will depend on growth conditions and rhizobial effectiveness, but it is agreed that on the question of the energetics of the actual fixation process Gibson's work is more informative. It is possible that useful data on this aspect could be gained by the simultaneous measurement of respiration and fixation in detached non-legume nodules, as attempted on a small scale by Bond and MacConnell (1955).

In further discussion Gibson (*loc. cit.*) contemplates that the carbohydrate consumption for fixation may actually be less than for the uptake of nitrogen in the combined form. However, arguing rather teleologically it would be difficult to reconcile this with the fact that both legume and non-legume nodulated plants have a mechanism which eventually suppresses fixation when ample supplies of combined nitrogen are available.

NON-LEGUMES WITH *ALNUS*-TYPE NODULES

Table II lists the genera concerned and provides brief information relating to them. Most of the estimates of the number of good species in the genera were kindly provided by Mr H. K. Airy Shaw, Royal Botanic Gardens, Kew, while the numbers of species recorded to bear nodules are based on a recent survey of the very scattered literature by C. Rodriguez-

Barrueco in the author's laboratory. The names of many of these species and the appropriate references were provided by Allen and Allen (1965). When the species so far unexamined have been duly inspected, many of these genera may prove to be wholly nodulating.

TABLE II

Non-legume genera with *Alnus*-type nodules.

Genus	Species complement	Distribution	Species recorded to bear nodules
Coriaria	15	Mediterranean, Central America, Japan, New Zealand	12
Alnus	35	Europe, Siberia, North America, Andes, Japan	25
*Myrica**	35	Tropical, sub-tropical and temperate regions	12
Casuarina	45	Australia, tropical Asia, Pacific Islands	14
Elaeagnus	45	Asia, Europe, North America	9
Hippophaë	1	Asia, Europe	1
Shepherdia	3	North America	2
Ceanothus	55	North America	30
Discaria	10	South America, New Zealand, Australia	1
Dryas	4	Arctic and north temperate zone	3
Purshia	2	North America	2
Cercocarpus	20	North America	1
Arctostaphylos	40	North-west America, Europe, Asia	1

* Including *Comptonia*.

Although efforts to isolate the endophytes from these nodules have continued to meet with no confirmed success, the use of the electron microscope for the study of the very congested contents of the infected cells of the nodules by Silver (1964), Becking *et al.* (1964) and Gardner (1965) has provided confirmation of the hyphal nature of the endophyte, the dimensions of the hyphae being similar to those of an actinomycete. This was actually a view which had been widely adopted on the basis of light microscopy, until Hawker and Fraymouth (1951) declared their inability to find true hyphae in the nodules.

In the author's laboratory Rodriguez-Barrueco has found that the *Alnus glutinosa* and *Myrica gale* organisms can be shown to be present in Scottish soils which have not supported the host plants for long periods of years. It is also becoming clear that the endophytes normally associated with different species of a given host genus, especially when the species are from

(a)

(b)

FIG. 1. (a) Nodulated plants of *Casuarina cunninghamiana* after seven months' growth in water culture without combined nitrogen, those on the left with, and those on the right without cobalt (× $\frac{1}{5}$). From Hewitt and Bond (*J. exp. Bot.* **17**, 480). (b) Non-nodulated plants of *Alnus glutinosa* after three months' growth in water culture with ammonium-nitrogen supplied, those on the left with, and those on the right without cobalt (× $\frac{1}{5}$). Hewitt and Bond (original)

(a)

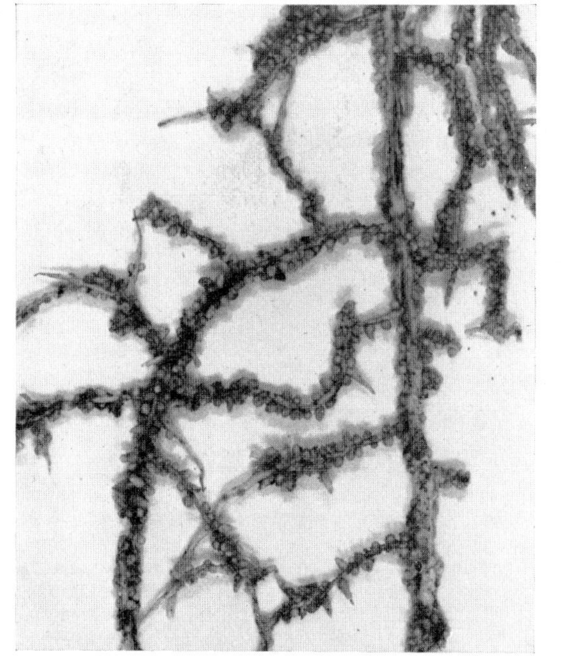

(b)

FIG. 2. (a) Nodule clusters on roots of *Encephalartos villosus*, from the Glasgow Botanic Garden (× $1\frac{1}{2}$); (b) Roots of *Podocarpus macrophyllus* with nodules, from the Royal Botanic Garden, Edinburgh (× $1\frac{1}{2}$).

The photographs were taken by Mr R. Cowper.

distinct geographical regions, are not necessarily fully identical. Thus Rodriguez-Barrueco (1966) found that the South American species, *Alnus iorullensis*, when inoculated from *A. glutinosa* nodules, produced unusually numerous, small nodules fixing little or no nitrogen. Gardner and Bond (1966) and Bond (1967) found the same features in African and American species of *Myrica* inoculated from *M. gale*. Becking (1966) reported defective symbiosis between the American species, *Alnus rubra*, and the *A. glutinosa* organism, while Rodriguez-Barrueco and Bond (1967) had a similar experience when three European species of *Alnus* were inoculated from *A. rubra* nodules.

Hewitt and Bond (see Bond, 1963 for references) demonstrated a molybdenum requirement for fixation in *Alnus*, *Myrica* and *Casuarina*, and subsequently showed that symptoms of intense nitrogen deficiency appeared in nodulated plants of the same genera when deprived of cobalt (Hewitt and Bond, 1966), as seen in Fig. 1a. This result was concluded to be mainly an outcome of interference with vitamin B_{12} synthesis in the nodular endophyte in the cobalt-deprived plants. No need for cobalt could be demonstrated in non-nodulated plants supplied with combined nitrogen (Fig. 1b).

Confirmation of the ecological importance of these plants is increasing. Thus there is the work of Lawrence (1953), and Crocker and Major (1955) showing the soil-enriching effects of *Dryas drummondii* and *Alnus crispa* on recently deglaciated areas in Alaska, and that of Dommergues (1963) showing the extent of nitrogen accumulation by *Casuarina* in African dunes. Stewart and Pearson (1967) have provided corresponding data for *Hippophaë* colonizing British dunes, while Goldman (1961) demonstrated the contribution of *Alnus* sp. to the fertility of fresh water, and Daly (1966) that of another *Alnus* sp. to the nitrogen status of forest soil. The circulation in the biosphere of nitrogen fixed by these plants is hastened by the relatively high nitrogen content of their leaves at leaf fall. Thus in unpublished data obtained by Rodriguez-Barrueco, while freshly fallen leaves of oak, lime and elm contained close to 1% nitrogen in the dry matter, those of alder growing in the same area contained 2·8%.

SYMBIOSES INVOLVING BLUE-GREEN ALGAE

These are undoubtedly attended by fixation. Thus the work of Bergersen *et al.* (1965) and of Bond (1967) show this for the nodules of the cycads *Macrozamia*, *Ceratozamia* and *Encephalartos* (Fig. 2a). The Australian study showed for the first time that the nitrogen fixed in the nodules is exported to other parts of the cycad; thus within a period of 48 hours ^{15}N was fixed in the nodules and translocated to petioles, stem and roots, though it had not reached the laminae in detectable amount.

The experiments of Bortels (1940), Fujiwara *et al.* (1947) and Saubert (1949) showed the ability of plants of the floating fern, *Azolla*, associated with *Anabaena*, to accumulate nitrogen and grow in a nitrogen-free medium. Saubert calculated that the fixation was equivalent to a gain of 312 kg nitrogen/hectare—a very large amount. It is recorded (*Use of leguminous plants in tropical countries as green manure, as cover and as shade*, International Institute of Agriculture, Rome (1936)) that in some regions there has been the custom of introducing *Azolla* into rice-fields in the cold season, since its growth had been found to increase fertility. It is also of interest that *Azolla* species were widespread in Europe during interglacial periods.

The *Nostoc* cavities in *Blasia*, one of the liverworts showing this feature, were found by Bond and Scott (1955) to be attended by nitrogen fixation. About one-fifth of lichen genera involve blue-green algae; Bond and Scott (*loc. cit.*) and Scott (1956) provided ^{15}N and other evidence that such lichens fix nitrogen, while those involving green algae do not.

LEAF-BACTERIA ASSOCIATIONS

The so-called leaf nodules, present in certain large genera of the families Rubiaceae and Myrsinaceae, consist of cavities within the leaf tissues occupied by dense growths of a bacterium. Species with leaf nodules are prominent in the floras of some regions of Africa, Madagascar, and other tropical and semi-tropical countries. The status of these nodules with regard to nitrogen fixation is not absolutely clear; their investigation is rather difficult, especially since the seeds carry the bacterium, and are not readily freed from it. Silver *et al.* (1963) and Centifanto and Silver (1964) have re-examined the nodules of *Psychotria*, and are satisfied that the bacterium is a species of *Klebsiella*. They showed that the isolate fixed nitrogen provided that conditions were anaerobic; it is just a little difficult to suppose that such a condition could exist in a photosynthetic organ. Hanada (1954) isolated a nitrogen-fixing bacterium, identified as *Xanthomonas*, from leaf nodules of *Ardisia*, and provided evidence that nodulated plants were able to grow in a nitrogen-free rooting medium.

The concept that the wet leaf surface—the phyllosphere—of vegetation in humid tropical regions might provide a site for microbial development of significance in the nutrition of the plant is largely due to Ruinen (1961, 1965), who found that *Beijerinckia* and *Azotobacter* were regularly present on the leaf surface, along with many other micro-organisms. She envisaged that these organisms were nourished by nutrients leached from the leaf, and that fixation proceeded, its products being absorbed by the leaf. The attractive feature of this concept is that it does not depend on the presence of special structures such as nodules, but could operate in many species under appropriate climatic conditions. Among supporting evidence gathered by Ruinen are data showing that dew and the run-off from leaves during rain from various species in Surinam contained measurable amounts of carbohydrates and amino acids, and increased in nitrogen content on standing. In the Netherlands, Ruinen tried to simulate rain forest conditions by floating detached leaves of *Phaseolus vulgaris* and *Coffea arabica* in petri dishes containing nitrogen-free culture solution, and leaving them in the greenhouse for some days. Subsequent analysis provided evidence of an increase in nitrogen content of the leaves, attributed to fixation by bacterial organisms originally present on the leaves and enabled to become active under the experimental conditions, nourished by carbohydrates leaching from the leaf. Inoculation with *Azotobacter* increased the gain in nitrogen. In passing, these leaf-culture tests seem to be significant in connection with the frequent use of leaf discs in the study of carbohydrate and nitrogen metabolism in leaves.

It is obvious that the phyllosphere concept is of considerable potential importance. It is to be hoped that further supporting evidence will be provided.

MYCORRHIZA

First will be considered the mycorrhizal condition present in the Podocarpaceae, a large and economically important gymnospermous family, and some related genera. The fungus is housed in small hemispherical nodules which occur in tremendous numbers along the roots (Fig. 2b). The recent observations of Baylis *et al.* (1963), Bergersen and Costin (1964), and the electron microscope study by Becking (1965) have removed earlier doubt concerning the identity of the endophyte, and confirm that it is a vesicular-arbuscular fungus. Baylis *et al.* (*loc. cit.*) see a close analogy between these fungal nodules and the coralloid ectotrophic mycorrhizal roots of *Pinus*, and with this in mind they compared the growth of infected and uninfected plants of *Podocarpus* spp. in a soil deficient in phosphorus and calcium, but not in nitrogen. Growth of the infected plants was much superior, attributed to more efficient uptake of the deficient nutrients. In contrast, Bergersen and Costin (*loc. cit.*) and Becking (*loc. cit.*), bearing in mind earlier suggestions that podocarp nodules might be nitrogen-fixing, tested this possibility, the first authors by a ^{15}N technique only, the last-named by plant culture as well. The isotopic method proved more profitable, and the outcome of these two studies was that the four samples of nodulated roots exposed to ^{15}N all showed significant enrichment, although the level of fixation indicated was low compared with other nodular material. Thus it appears that the fungal association brings two-fold advantage to the podocarp host plant.

There is a dearth of investigations into the possible association of fixation with other examples of mycorrhiza. As with the podocarps, some workers have arranged their experiments so as to test for a promotion of mineral uptake by the fungal associate (Morrison, 1957; Baylis, 1962). Stevenson (1959) provided some evidence of fixation in *Pinus radiata* mycorrhiza, and Richards and Voigt (1964) sought to extend this under aseptic conditions, but failed in this difficult undertaking. Among other possibilities which they considered, Hassouna and Wareing (1964) suggested that mycorrhiza might have been responsible for a fixation of nitrogen which they reported to be associated with marram grass (*Ammophila* sp.). This was based chiefly on the finding that when plants of this species were grown in dune sand in pots for two years, the nitrogen content of plant and sand together increased by nearly 200 mg on average, while unplanted pots showed an increase of only 16 mg. It would have been useful if some indication of the variation among replicate pots could have been included, while the use of deionized rather than distilled water involves a slight risk in work of this kind. However, the ability of *Ammophila* to grow strongly in dune sand of low nitrogen content is clearly a matter of interest.

Some of the topics considered in this paper receive more detailed treatment in a forthcoming review (Bond, 1967).

Since presenting this paper, the account by Yamada (1960) of his re-investigation of the leaf nodules of *Ardisia* has been seen. Cultures of the organism (named as *Bacterium foliicola*) and also nodulated leaves gave negative results when tested for fixation by ^{15}N; the gas mixture employed contained 40% of oxygen.

REFERENCES

Allen, E. K. and Allen, O. N. (1965) *In* "Microbiology and Soil Fertility" (Gilmour, C. M. and Allen, O. N. eds.) pp. 77–106. Proc. 25th Ann. Biol. Colloq., Oregon State University Press.

Baylis, G. T. S. (1962). *Aust. J. Sci.* **25**, 195.

Baylis, G. T. S., McNabb, R. F. R. and Morrison, T. M. (1963). *Trans. Br. mycol. Soc.* **46**, 378.

Becking, J. H. (1965). *Pl. Soil* **23**, 213.

Becking, J. H. (1966). *Annls. Inst. Pasteur, Paris*, **111**, 211.

Becking, J. H., De Boer, W. E. and Houwink, A. L. (1964). *Antonie van Leeuwenhoek*, **30**, 343.

Bergersen, F. J. and Costin, A. B. (1964). *Aust. J. biol. Sci.* **17**, 44.

Bergersen, F. J. (1965). *Biochim. biophys. Acta* **115**, 247.

Bergersen, F. J., Kennedy, G. S. and Wittman, W. (1965). *Aust. J. biol. Sci.* **18**, 1135.

Bond, G. (1941). *Ann. Bot.* **5**, 313.

Bond, G. (1963). *In* "Symbiotic Associations" (Nutman, P. S. and Mosse, B. eds.) pp. 72–91. Soc. gen. Microbiol., Symp. 13, Cambridge University Press, London.
Bond, G. (1967). *A. Rev. Pl. Physiol.* **18**, *107.*
Bond, G. (1967). *Phyton B. Aires* **24,** 57.
Bond, G. and MacConnell, J. T. (1955). *Nature, Lond.* **176**, 606.
Bond, G. and Scott, G. D. (1955). *Ann. Bot.* **19**, 67.
Bortels, H. (1940). *Arch. Mikrobiol.* **11**, 155.
Burris, R. H. (1966). *A. Rev. Pl. Physiol.* **17**, 155.
Centifanto, Y. M. and Silver, W. S. (1964). *J. Bact.* **88**, 776.
Crocker, R. L. and Major, J. (1955). *J. Ecol.* **43**, 427.
Daly, G. T. (1966). *Can. J. Bot.* **44**, 1607.
Dommergues, Y. (1963). *Agrochimica* **7**, 335.
Fujiwara, A., Tsuboi, I. and Yoshida, F. (1947). *Nogaku* (*Sci. of Agric.*) **1**, 361.
Gardner, I. C. (1965). *Arch. Mikrobiol.* **51**, 365.
Gardner, I. C. and Bond, G. (1966). *Naturwissenschaften* **53**, 161.
Gibson, A. H. (1966). *Aust. J. biol. Sci.* **19**, 499.
Goldman, C. R. (1961). *Ecology* **42**, 282.
Hanada, K. (1954). *Jap. J. Bot.* **14**, 235.
Hassouna, M. G. and Wareing, P. F. (1964). *Nature, Lond.* **202**, 467.
Hawker, L. E. and Fraymouth, J. (1951). *J. gen. Microbiol.* **5**, 369.
Hewitt, E. J. and Bond, G. (1966). *J. exp. Bot.* **17**, 480.
Lawrence, D. B. (1953). "Development of vegetation and soil on deglaciated terrain of Southeastern Alaska with special reference to the accumulation of nitrogen". Final Report, Off. Naval Res. Project No. 160-83, Washington.
Morrison, T. M. (1957). *New Phytol.* **56**, 247.
Pate, J. S. and Wallace, W. (1964). *Ann. Bot.* **28**, 83.
Pate, J. S., Walker, J. and Wallace, W. (1965). *Ann. Bot.* **29**, 475.
Richards, B. N. and Voigt, G. K. (1964). *Nature, Lond.* **201**, 310.
Rodriguez-Barrueco, C. (1966). *Phyton, B. Aires* **23**, 103.
Rodriguez-Barrueco, C. and Bond, G. (1967). *Proc. 40th Annual Mtg. of the Northwest Scientific Association*, Pullman Washington. In press.
Ruinen, J. (1961). *Pl. Soil* **15**, 81.
Ruinen, J. (1965). *Pl. Soil* **22**, 375.
Saubert, G. G. P. (1949). *Ann. bot. Gdn Buitenz.* **51**, 177.
Scott, G. D. (1956). *New Phytol.* **55**, 111.
Silver, W. S. (1964). *J. Bact.* **87**, 416.
Silver, W. S., Centifanto, Y. M. and Nicholas, D. J. D. (1963). *Nature, Lond.* **199**, 396.
Stevenson, G. (1959). *Ann. Bot.* **23**, 622.
Stewart, W. D. P. (1966). *In* "Nitrogen Fixation in Plants". The Athlone Press, London.
Stewart, W. D. P. and Pearson, M. C. (1967). *Pl. Soil* **26**, 348.
Turchin, F. V., Berseneva, N. and Zhidkikh, C. G. (1963). *Dokl. Akad. Nauk SSSR, Biological Sciences Sections* **149**, 522.
Yamada, T. (1960). *Bull. Faculty Educatn. Chiba Univ. Japan* **9**, 1.

Section 1c

Nitrogen Fixation in the Blue-Green Alga *Anabaena cylindrica*

ROSALIE M. COX

Department of Botany
University of Bristol, England

INTRODUCTION

In recent years, work with cell-free extracts of nitrogen-fixing bacteria has provided much information on the ancillary reactions involved in nitrogen reduction. The main requirements for enzymic nitrogen reduction have been shown to be a source of reducing power and energy in the form of ATP. A key redox compound involved in nitrogen reduction is ferredoxin (Mortenson, 1964) and other still uncharacterized redox compounds may transport electrons from ferredoxin to the nitrogen-activating enzyme. The step or steps which utilize ATP are not certain but recent suggestions are that ATP is required to activate reduced cofactors (Dilworth *et al.*, 1965; Hardy and Knight, 1966). Ferredoxin can be reduced directly by gaseous hydrogen in the bacterium *Clostridium pasteurianum* (Mortenson, 1964) but the outstanding organic compound demonstrated to donate electrons to ferredoxin is pyruvate. The phosphoroclastic reaction of pyruvate involves the elimination of two electrons and carbon dioxide and the phosphorylation of the remaining acetyl moiety (Carnahan *et al.*, 1960). In the presence of ADP, ATP may be generated from acetyl phosphate so pyruvate may supply both basic requirements of nitrogen reduction. The reactions described above are summarized in Fig. 1.

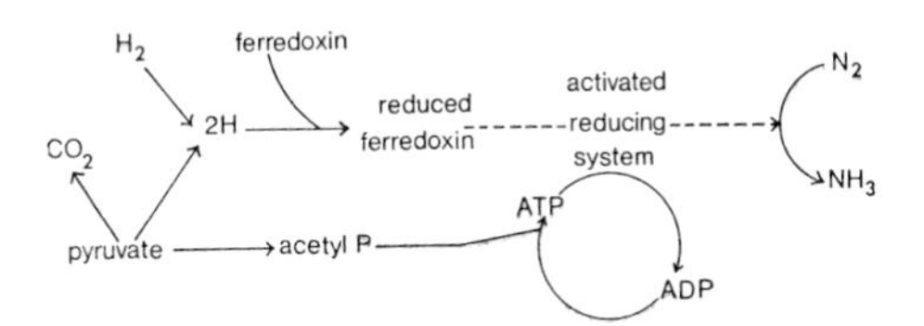

FIG. 1. Scheme of the reactions suggested to be involved in N_2 fixation. (From Cox, *Arch. Mikrobiol* **56,** 193)

In photosynthetic nitrogen-fixing organisms, the possibility exists that reduced ferredoxin and/or ATP generated photochemically may be directly used for nitrogen fixation. It is therefore of interest to know whether these organisms can in fact utilize light-generated factors for nitrogen reduction or whether, in such organisms, the same pathway of electron transport operative in heterotrophic nitrogen fixers exists.

In long-term experiments, nitrogen fixation in photosynthetic organisms is usually dependent upon light and a supply of carbon dioxide (Fogg and Than-Tun, 1960; Fay and Fogg, 1962) although certain photosynthetic nitrogen fixers can grow and fix nitrogen in the dark using organic compounds (Allison *et al.*, 1937; Fay, 1965). One obvious dependency of nitrogen fixation upon light is for a source of carbon skeletons to assimilate the fixed nitrogen. There have, however, also been suggestions that the requirement of reducing power and energy may also be supplied photochemically. Such suggestions have been made for *Chromatium* sp. (Arnon *et al.*, 1961), *Rhodospirillum rubrum* (Pratt and Frenkel, 1959) *Chlorogloea fritschii* (Fay and Fogg, 1962), *Anabaena cylindrica* (Fogg and Than-Tun, 1958; Cobb and Myers, 1964) and *A. flos-aquae* (Davis and Tischer, 1966). Fogg and Than-Tun (1958) found that more oxygen was evolved in the light under nitrogen-fixing conditions than by cells not fixing nitrogen. The ratio of the amount of nitrogen fixed to amount of additional oxygen evolved was about 1 : 1·5. This ratio is consistent with the overall formula:

$$N_2 + 3H_2O = 2NH_3 + 1{\cdot}5\ O_2$$

when it is assumed that reducing power for nitrogen fixation is generated from water photochemically. However, Than-Tun and Harris (quoted by Fogg, 1961) found that after 12 h illumination *A. cylindrica* could fix nitrogen also in the dark. The design of this experiment permits no conclusions as to whether nitrogen fixation was supported by dark-generated factors or by factors produced photochemically during the preceding light period.

For studying the relations between nitrogen fixation and light in short-term experiments with intact cells of the blue-green alga *A. cylindrica*, manometric methods for measuring relevant gas exchanges have been developed. These methods and some of the results obtained using manometry are described below.

METHODS

Culture methods. Bacteria-free cultures of *A. cylindrica* (Cambridge Culture Collection No. 1403/2) were grown in bulk in nitrogen-free mineral medium (Allen and Arnon, 1955). The growth vessels were shaken continuously to keep the algal filaments in an even suspension and to allow more rapid gas exchange between gas and liquid phases. Continuous illumination of 350 foot-candles was provided and the temperature maintained at 25° or 30°C.

Exponentially-growing cultures were harvested and resuspended in 50 ml fresh sterile medium. Nitrogen starvation was carried out by incubating the resuspended cells under 2% carbon dioxide in argon for 12 or 16 h with continuous shaking at a light intensity of 600 foot-candles.

Manometry. Samples of algal suspension (3 or 4 ml) were dispensed in Warburg flasks and gas exchange measured by manometry. Flasks were gassed with nitrogen or argon before experiments. Oxygen evolution in the light was taken to be a measure of photosynthesis, carbon dioxide (0·2% in the gas phase) being supplied by 0·4 ml of a buffer (Pardee, 1949) in the centre well. Nitrogen uptake in the light was measured by adding 0·5 ml $CrCl_2$ (Myers, 1960) to the side-arm of appropriate flasks to absorb photochemically-produced oxygen. The quantity of nitrogen fixed in the light was added to the net quantity of oxygen evolved under identical conditions to calculate gross oxygen evolution. Dark flasks were provided with 0·5 ml of 10% KOH solution to absorb any carbon dioxide evolved. Experiments were carried out at 30°C and at a light intensity of 500 or 650 foot-candles.

The dry weight of aliquots of experimental algal suspensions was determined so that the results of all experiments could be evaluated on a common basis.

Gravimetric methods and radioactivity assays. For estimation of carbon dioxide absorbed in potassium hydroxide at the end of manometric measurements, the alkali was quantitatively removed from the side-arm and absorbed carbon dioxide converted to barium carbonate. The precipitate was filtered on a weighed membrane filter, dried and reweighed. Radioactive barium carbonate precipitates were counted using a Geiger counter and also weighed in order to correct for self-absorption of the radioactivity.

Further experimental details may be found in earlier papers (Cox, 1966, 1967).

RESULTS

RELATIONSHIP BETWEEN NITROGEN UPTAKE AND CARBON DIOXIDE FIXATION

Nitrogen fixation can be measured with reasonable facility by manometry although a fairly large quantity of algal material (10–15 mg dry weight/flask) is needed to give appreciable nitrogen uptake. The rate of nitrogen uptake during experimentation may be increased by a previous period of nitrogen starvation (Fig. 2) during which the nitrogen content of the cells drops on a dry weight basis (Table I). Nitrogen uptake shown in Fig. 2 was measured

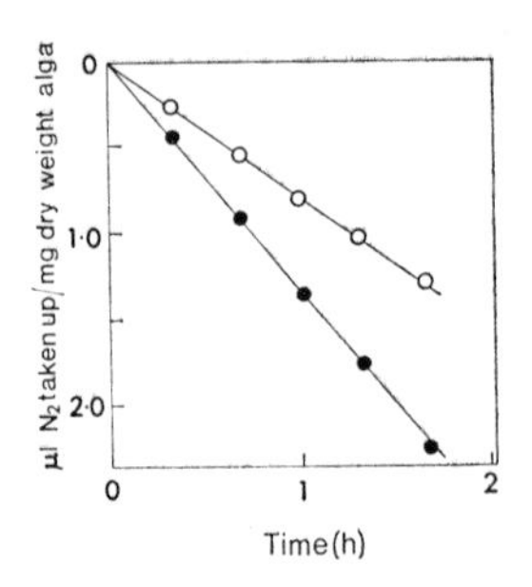

FIG. 2. Nitrogen fixation in the light by N_2-starved and normal algal suspension. Gas uptake by 4 ml samples was measured in N_2-gassed flasks provided with CO_2. Light intensity, 650 foot-candles. ○, normal, ●, N_2-starved. (From Cox, *Arch. Mikrobiol.* **53**, 263)

TABLE I

Percentage N_2 in normal and N_2-starved material.

Algal material grown for 6 days was divided into two equal portions and each concentrated into 50 ml sterile medium. One portion was N_2 starved and the other allowed to fix N_2. Dry weights were obtained by drying aliquots (3ml) of material at 110°C overnight in weighed crucibles. Similar algal aliquots were digested by a Kjeldahl procedure and total N_2 estimated spectrophotometrically by a standard Nessler method.

Sample	Dry weight (mg)	Total N_2 (mg)	Total N_2 as % of dry weight
Normal cells	20·2	1·57	7·8
N_2-starved cells	18·7	1·20	6·4

(From Cox, *Arch. Mikrobiol.* **56**, 193)

in the presence of added carbon dioxide in the light and shows a steady rate of uptake during 2 h incubation. In the absence of carbon dioxide in the light, a different pattern of gas uptake was observed (Fig. 3). An initial phase of rapid gas uptake was followed by a progressively decreasing rate. For comparison, gas uptake in argon was also measured and proved to be negligible when compared with gas uptake in nitrogen showing the latter to be entirely due to nitrogen fixation. That gas uptake was observed at all in argon was probably due to some nitrogen impurity. The pattern of gas output was closely parallel to that of nitrogen uptake and the final quantities of gas exchanged give a ratio of nitrogen fixed to oxygen evolved of 1 : 1·4,

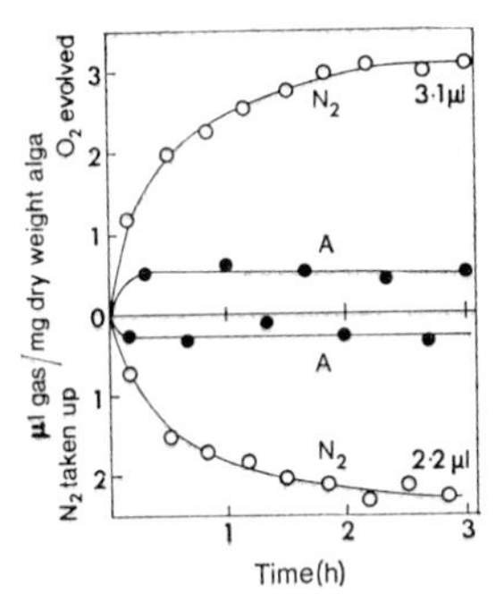

FIG. 3. Gas exchange by N_2-starved cells in the light in the absence of CO_2. Gas exchange by 4 ml samples was measured in flasks gassed with N_2 or argon (A). Total quantities of gas exchanged are indicated on the graph. Light intensity, 650 foot-candles. (From Cox, *Arch. Mikrobiol.* **56**, 193)

near that reported by Fogg and Than-Tun (1958), Cobb and Myers (1964) and Davis and Tischer (1966). It is probably significant that the value of this ratio is not invariable from such experiments, having been observed to range from 1 : 0·9 to 1 : 3·0.

When gas uptake was measured in the dark, a similar pattern to that shown in Fig. 3 was observed (Fig. 4). Again, gas uptake in argon was

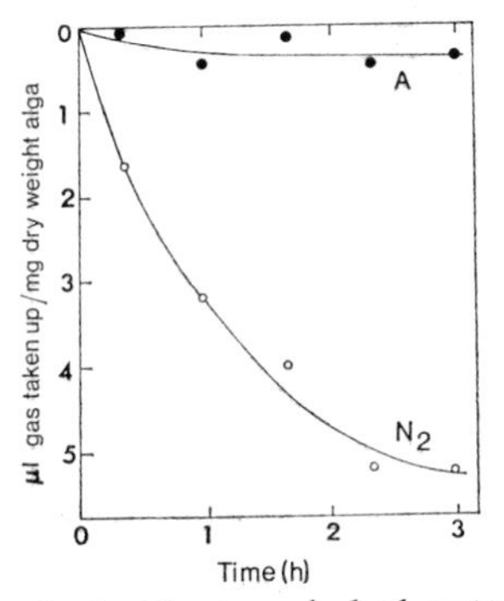

FIG. 4. Gas uptake by N_2-starved algal material in the dark. Gas uptake by 4 ml samples was measured in the dark in flasks gassed with N_2 or argon. Alkali was present in the side-arm of each flask to absorb CO_2.

negligible while gas uptake in nitrogen proceeded first at a fast rate followed by a progressively decreasing rate. Evidently in nitrogen-starved cells nitrogen fixation may occur in the dark. Studies on $^{15}N_2$ uptake by intact cells incubated in the dark have confirmed that nitrogen fixation can take place in the dark (Cox and Fay, 1967). It is improbable that light-generated cofactors could have been supporting dark nitrogen fixation as experimental measurements were made after gassing samples for one hour in the dark.

A more likely explanation for the occurrence of dark nitrogen fixation is that some dark process was providing necessary cofactors. It may be noted that no nitrogen uptake in the dark was observable unless potassium hydroxide was present indicating that carbon dioxide evolution is a concomitant of dark-nitrogen uptake.

The pattern of gas uptake obtained from cells incubated in the light without carbon dioxide (Fig. 3) and in the dark (Fig. 4) may be interpreted as the initial availability of a limited amount of suitable carbon skeletons to assimilate the ammonia produced by nitrogen fixation, followed by the slow conversion of accumulated carbohydrate to such carbon skeletons suitable for amination. So far, no information is available concerning the nature of the carbohydrate accumulated during nitrogen starvation.

To test the theory that the differing pattern of nitrogen uptake in the light in the presence of carbon dioxide (Fig. 2) and in the light without carbon dioxide (Fig. 3) or in the dark (Fig. 4) was due largely to the availability of carbon skeletons, nitrogen uptake in the light with and without carbon dioxide was compared in the same experiment. Oxygen output under these two conditions was also measured and a set of argon-gassed control flasks was included (Fig. 5). The same patterns of nitrogen uptake shown in Figs.

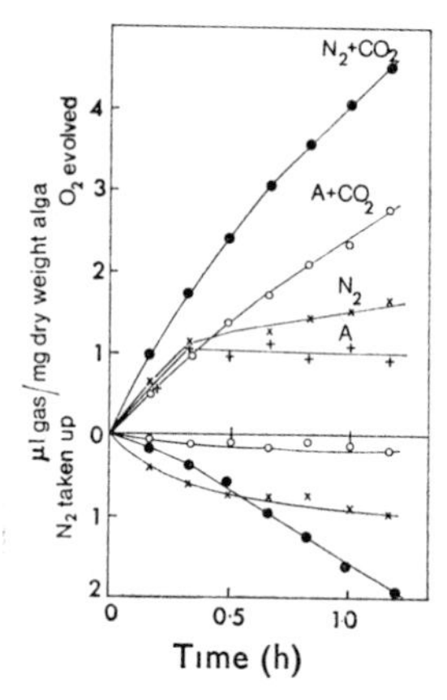

FIG. 5. Effect of CO_2 on gas exchange by N_2-starved cells in the light. Samples (4 ml) were gassed with N_2 or argon and N_2 uptake or O_2 output measured in the light (650 foot-candles) with or without 0·4 ml CO_2 buffer. The symbols used for O_2 evolution under different conditions pertain to equivalent conditions when used for N_2 uptake.

2 and 3 were repeated in this experiment. A constant rate of nitrogen fixation was obtained in the presence of carbon dioxide and a diminishing rate of nitrogen uptake was observed in the sample lacking carbon dioxide. Again, gas uptake by argon-gassed samples was negligible, none at all being detected from the sample lacking carbon dioxide. There seems no doubt that

the gas uptake measured manometrically was due to nitrogen fixation and that, for a constant rate of nitrogen uptake, a continuous supply of carbon skeletons, preferably provided directly by photosynthesis, is required.

An interesting effect of nitrogen fixation upon oxygen evolution is shown also in Fig. 5. It is evident that, under nitrogen-fixing conditions, oxygen evolution proceeded at a faster rate than when nitrogen fixation was prevented. This observation is the same as that made by Fogg and Than-Tun (1958) using different experimental methods. Their interpretation of the extra oxygen evolution being due to participation of photochemical systems in nitrogen reduction was based on their finding that the extra oxygen evolution could not be accounted for by extra carbon dioxide assimilation. Unfortunately, to measure the rate of carbon dioxide fixation in the absence of nitrogen uptake, ammonia, a known uncoupler of phosphorylation, was used and this probably accounts for the partial inhibition of carbon dioxide fixation which they observed. In an experiment in which $^{14}CO_2$ fixation by argon-gassed cells was compared to that by cells fixing nitrogen (Cox, 1966), it was found that carbon dioxide fixation was, in fact, more rapid in the latter material. The increase in carbon dioxide assimilation was of the same order as the increase in rate of oxygen evolution in nitrogen-fixing cells. The observation of extra oxygen evolution in the light under nitrogen-fixing conditions is not therefore definite evidence of light supplying photochemical products for use in nitrogen fixation.

In a later paper, Fogg and Than-Tun (1960) demonstrated a close dependency of carbon dioxide fixation upon partial pressure of nitrogen and it is therefore likely that the extra oxygen evolution observed under nitrogen-fixing conditions is due mainly to stimulation of carbon dioxide assimilation. Carbon dioxide fixation is probably controlled by the availability of ammonia to aminate some of the early products of photosynthesis to maintain the carbon/nitrogen balance of the cell. In cells fixing nitrogen as the sole nitrogen source, the rate of nitrogen fixation would be expected to control the rate of carbon dioxide assimilation. The two processes are therefore interdependent.

NITROGEN FIXATION AND PYRUVATE METABOLISM

Since nitrogen fixation appeared to be dependent on light mainly for a supply of carbon skeletons and could be supported for a short period by endogenous carbon skeletons, a number of organic compounds were supplied to nitrogen-fixing cells as possible alternative carbon skeletons. It was thought that such an alternative would facilitate easier study of the relationship between nitrogen fixation and light in the absence of the complication of carbon supply. The compounds chosen were phosphoglycerate, an early product of carbon dioxide fixation; α-oxoglutarate, since glutamic acid is the first organic product of nitrogen fixation in most nitrogen fixers; fumarate and succinate, which act as carbon skeletons for nitrate assimilation in *A. cylindrica* (Wolfe, 1954), and pyruvate, known to supply cofactors for nitrogen reduction. These compounds were supplied to nitrogen-starved cells incubated in the light in the absence of carbon dioxide. A control flask lacking organic substrate was supplied with carbon dioxide so that the rate of nitrogen fixation using the natural carbon source could be compared. The result of this experiment is shown in Fig. 6.

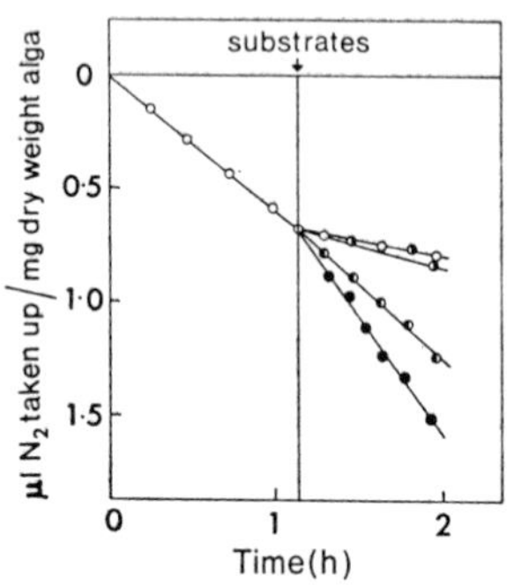

FIG. 6. Uptake of N_2 in the light by N_2-starved cells supplied with CO_2 or organic substances in the absence of CO_2.
Nitrogen-gassed samples (4 ml) were incubated in the light (650 foot-candles). After a constant rate of N_2 uptake was obtained, the substances indicated, each as sodium salts, (10 μmoles in 0·2 ml) were tipped into the samples and the new rate of N_2 uptake measured. ○, α-oxoglutarate; ◑, fumarate; ◐, CO_2; ●, pyruvate.

No gas uptake could be detected in the presence of succinate or phosphoglycerate and negligible gas uptake was recorded on the addition of α-oxoglutarate and fumarate. Pyruvate, however, markedly stimulated gas uptake to a rate which exceeded that supported by photosynthetically-produced carbon skeletons. Dark-nitrogen uptake was also stimulated by pyruvate (Fig. 7). Metabolism of pyruvate by an oxidative process was

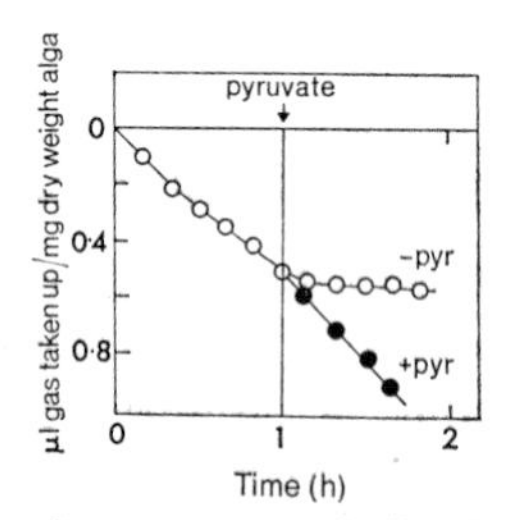

FIG. 7. Effect of pyruvate on N_2 fixation in the dark.
Nitrogen-starved samples (4 ml) were gassed with N_2 and sodium pyruvate (10 μmoles in 0·2 ml) added when a constant rate of N_2 uptake had been obtained. (From Cox, *Arch. Mikrobiol.* **53**, 263)

excluded in this experiment by using anaerobic conditions. Owing to the known role of pyruvate as a generator of cofactors for nitrogen reduction in bacteria, no conclusion could be drawn from the results in Figs. 6 and 7 as to whether pyruvate was fulfilling the same function in *A. cylindrica* or acting as a carbon skeleton.

The action of pyruvate on nitrogen uptake in the presence or absence of carbon skeletons provided by photosynthesis is demonstrated in Fig. 8. Stimulation of nitrogen fixation by pyruvate was most marked in the presence of abundant carbon skeletons from photosynthesis and it therefore appears unlikely that the primary role of pyruvate in nitrogen reduction is to provide carbon skeletons. If pyruvate were being used as carbon skeletons preferentially to those provided by carbon dioxide assimilation, much more marked stimulation of nitrogen uptake would be expected in those cells lacking a supply of carbon dioxide. This is not the case, as shown in Fig. 8,

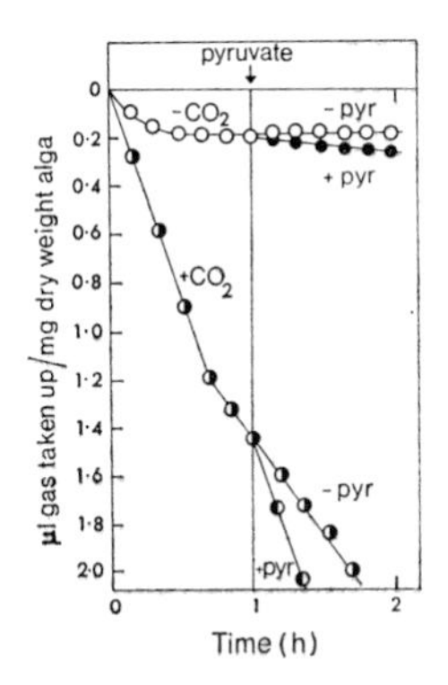

FIG. 8. Effect of CO_2 assimilation on the stimulation of N_2 fixation by pyruvate. Nitrogen-gassed samples (4 ml) of N_2-starved cells were incubated in the light (650 foot-candles) either in the presence or absence of 0·4 ml CO_2 buffer. After a steady rate of N_2 uptake was obtained, sodium pyruvate (10 μmoles in 0·2 ml) was tipped into the samples and the new rate of gas uptake determined. (From Cox, *Arch. Mikrobiol.* **53**, 263)

so it is more probable that pyruvate supplies cofactors for nitrogen reduction in *A. cylindrica*.

According to the scheme of the involvement of pyruvate in nitrogen fixation by bacteria (Fig. 1), pyruvate is decarboxylated as a necessary concomitant of acting as an electron donor. Therefore, if pyruvate acts as an electron donor for nitrogen reduction in *A. cylindrica*, decarboxylation of pyruvate would be expected to be faster in nitrogen-starved cells, which fix nitrogen more rapidly, than in normal cells. Radioactive carboxyl-labelled pyruvate was supplied to samples of normal and nitrogen-starved cells which were incubated in the light or dark. Carbon dioxide evolved by the cells was converted to barium carbonate and counted (Table II). Negligible radioactivity was recovered from the light-incubated samples,

TABLE II

Radioactive CO_2 evolved from sodium pyruvate-1-^{14}C by normal and N_2-starved cells.

Samples (3 ml) were gassed with N_2 and incubated for 1·5 h in the light (500 foot-candles) or in the dark, with the addition of 400 μmoles pyruvate containing 1 μc radioactivity/sample. Evolved CO_2 was trapped in alkali in the flask side-arm and converted to $BaCO_3$ which was filtered, dried, weighed and counted.

Treatment	Counts/min in $BaCO_3$/mg dry weight alga	
	Normal cells	N_2-starved cells
Light $-CrCl_2$	0·25	0·21
$+CrCl_2$	0·10	0·11
Dark	4·95	20·25

(From Cox, *Arch. Mikrobiol.* **56**, 193)

probably because any carbon dioxide evolved was immediately reassimilated. Appreciable radioactivity was recovered from the dark-incubated flasks and that from the nitrogen-starved cells was measurably higher than from the normal cells. Manometric measurements made using the same samples as used for Table II, demonstrate that nitrogen uptake proceeded at a faster rate in nitrogen-starved cells than normal ones (Fig. 9). There is,

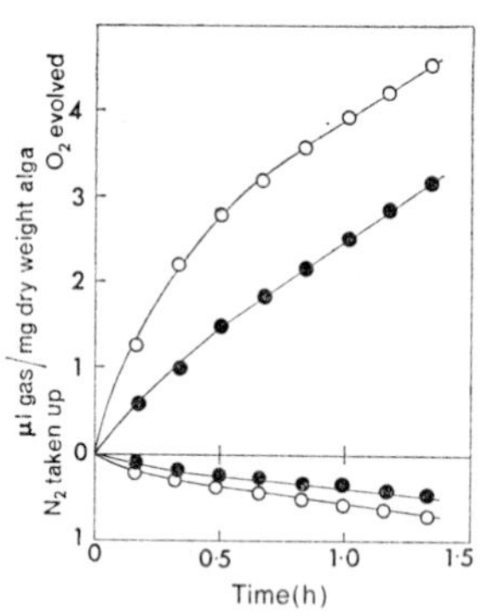

FIG. 9. Gas exchange by normal and N_2-starved cells in the light in the presence of pyruvate.
Nitrogen-gassed samples (3 ml) were incubated in the light (500 foot-candles) with the addition of 400 μmoles sodium pyruvate. The same symbols are used to indicate O_2 evolution and N_2 uptake by the two different materials. ○, nitrogen starved; ●, normal. (From Cox, *Arch. Mikrobiol.* **56,** 193)

therefore, a correlation between the rate of nitrogen fixed and the amount of pyruvate decarboxylated in *A. cylindrica* which may be taken to be evidence of pyruvate acting as a source of a cofactor or cofactors for nitrogen reduction in this organism.

It may be noted in Fig. 9 that, rather than obtaining the pattern of gas output observed with nitrogen-starved cells lacking carbon dioxide in Fig. 3, a fairly constant and appreciable rate of oxygen output from starved cells was measured after the initial rapid rate of gas evolution. Similarly, a fairly linear gas output was obtained from normal cells in the experiment. Taken together with the carbon dioxide analyses in Table II, the conclusion drawn from these observations is that carbon dioxide evolved from pyruvate may be re-assimilated in the light. Hence pyruvate could be active as a carbon skeleton in the light, albeit indirectly, in addition to its role as a generator of cofactors. If this is so, the observation of oxygen evolution during nitrogen fixation in the light in the absence of added carbon dioxide (Fogg and Than-Tun, 1958, and Figs. 3 and 5) could merely be due to the utilization of photochemically-produced metabolites in re-assimilating the carbon dioxide produced by decarboxylation. When pyruvate is acting as an electron donor for nitrogen fixation, nitrogen is, in effect, a terminal electron acceptor, no oxygen being consumed.

In order to establish whether pyruvate indirectly causes oxygen evolution by the method described above, oxygen output in the light by normal cells supplied with pyruvate was compared to that by cells lacking pyruvate (Fig. 10). Oxygen evolution from the samples provided with pyruvate

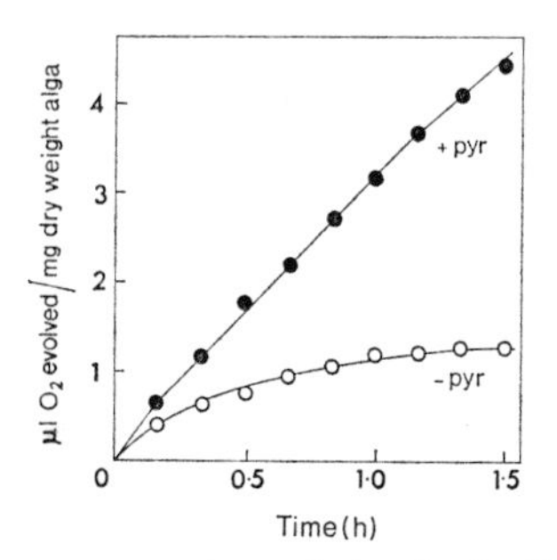

FIG. 10. Effect of pyruvate on O_2 evolution in the light by normal cells. Oxygen evolution from N_2-gassed samples (3 ml) was measured in the light (500 foot-candles) in the presence or absence of 400 μmoles sodium pyruvate. No CO_2 buffer was present. (From Cox, *Arch. Mikrobiol.* **56**, 193)

proceeded at an appreciable linear rate during the course of the experiment. The samples lacking pyruvate evolved a small amount of oxygen presumably by fixing carbon dioxide evolved from endogenous reserves, but at a decreasing rate throughout the experiment. Comparison of the quantities of carbon dioxide evolved in the light by the samples used in Fig. 10 and in similar dark-incubated samples, shows that a correspondingly larger amount of carbon dioxide was evolved from the cells provided with pyruvate (Table III). As in Table II, a much smaller amount of carbon dioxide was recoverable from the light-incubated samples.

TABLE III

Effect of pyruvate on CO_2 evolution from normal cells.

Nitrogen gassed samples (3 ml) were incubated for 1·5 h in the light (500 foot-candles) or dark. Evolved CO_2 was absorbed by alkali in the flask side-arm and converted to $BaCO_3$ which was filtered, dried and weighed.

Treatment	mg $BaCO_3$/mg dry weight alga	
	No pyruvate	400 μmoles pyruvate
Light	15·2	15·2
Dark	53·4	81·2

(From Cox, *Arch. Mikrobiol.* **56**, 193)

A further experiment to consider for later discussion of the significance of interrelations between photosynthesis, nitrogen fixation and pyruvate metabolism is the determination of the optimal light intensity for photosynthesis and nitrogen fixation. Figure 11 shows that, although the optimal light intensity for both processes is identical, the response of photosynthesis

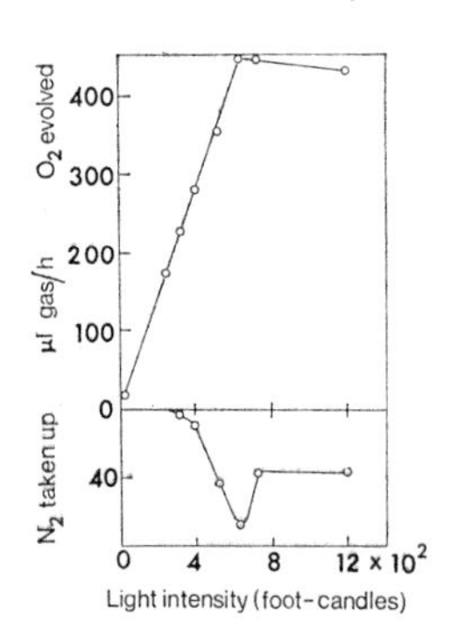

FIG. 11. Effect of light intensity on the rate of O_2 evolution and N_2 uptake by N_2-starved algal material.
Gas exchange by N_2-gassed samples (4 ml) provided with 0·4 ml CO_2 buffer was measured over a range of light intensities. The rates of gas exchange at each different light intensity have been calculated.

and nitrogen fixation to supra-optimal light intensity differs markedly. Whereas the rate of oxygen evolution was decreased only slightly at light intensities above optimal, nitrogen fixation was decreased to half its optimal rate. This short-term experiment is in agreement with a similar experiment of Cobb and Myers (1964) and the long-term experiments of Fogg and Than-Tun (1960).

DISCUSSION

From the experiments described, it is evident that the chief dependency of nitrogen fixation upon light in intact cells of *A. cylindrica* is for a continuous supply of carbon skeletons to assimilate ammonia, the first free product of nitrogen reduction. It has been demonstrated here that no reliable evidence yet exists for the further involvement of light in nitrogen fixation by supplying reducing power or ATP. On the other hand, pyruvate metabolism appears to play a role in nitrogen reduction and there is information in favour of the phosphoroclastic cleavage of pyruvate being able to supply cofactors for nitrogen fixation both in the light and dark. In cell-free extracts of *A. cylindrica*, a close correlation has also been observed between pyruvate decarboxylation and nitrogen fixation (Cox and Fay, 1967), both of these reactions being sited on the pigmented lamellae of the cell (Cox *et al.*, 1964; Fay and Cox, 1966). Similarly, the phosphoroclastic cleavage of pyruvate has been demonstrated to stimulate nitrogen fixation in cell-free extracts of two photosynthetic nitrogen-fixing bacteria, *Rhodospirillum rubrum* (Burris and Wang, 1960) and *Chromatium* sp. (Bennett *et al.*, 1964). Nitrogen fixation in certain autotrophic organisms, including *A. cylindrica*, can therefore be a dark process, when carbon skeletons are available, and is similar, under these circumstances, to the nitrogen-fixing system operative in heterotrophic bacteria.

It is reasonable to hope that some concrete evidence for the involvement of photochemical cofactors in nitrogen reduction may be forthcoming in the future in view of the finding that the enzymes for both oxygen evolution and nitrogen fixation are located on the pigmented membrane fraction of the cell (Cox *et al.*, 1964). Should future research demonstrate that photochemical products are involved in nitrogen fixation, it may be questioned whether the pyruvate system could be of use in addition to a photochemical system. Obviously in nitrogen-starved cells, pyruvate may play a role in nitrogen reduction since a supply of endogenous carbon skeletons is available. But would a physiological situation akin to nitrogen starvation arise under natural conditions? The results in Fig. 11 show that nitrogen starvation by natural means is potentially possible. Natural light intensities far in excess of those used in Fig. 11 are quite normal and would undoubtedly cause nitrogen starvation if supplied continuously to an algal suspension relying on gaseous nitrogen as its sole source. However, in natural environments light is intermittent and in the dark periods carbon reserves accumulated at high light intensity could be mobilized for ammonia assimilation, with pyruvate providing the necessary cofactors for nitrogen reduction. During the night, the appropriate carbon-nitrogen balance for optimal growth might thus be restored.

Nitrate reduction, in many ways an analogous process to nitrogen fixation, may use reducing power generated either photochemically or by respiration in certain photosynthetic organisms (Kessler, 1964). Therefore, especially since the electron transport functions of *A. cylindrica* all appear to be located on the same membranes, it is considered likely that nitrogen fixation may be as versatile as nitrate reduction, using either dark or light generated reducing power and ATP according to prevailing circumstances, unless spatial organization of the necessary enzymes prevents this.

SUMMARY

Short-term manometric experiments with the blue–green alga *Anabaena cylindrica* show that the close dependency of nitrogen fixation upon photosynthesis can be eliminated temporarily in nitrogen-starved cells. Nitrogen uptake by such cells occurred in the light in the absence of carbon dioxide and in the dark but at a progressively decreasing rate. Continued steady nitrogen uptake was only maintained for long periods in the presence of carbon dioxide in the light. In the dark, nitrogen uptake was accompanied by carbon dioxide evolution.

Oxygen output in the light both with and without carbon dioxide was stimulated under nitrogen-fixing conditions. In the presence of added carbon dioxide the stimulation of oxygen evolution may be accounted for by additional carbon dioxide assimilation.

Of a number of organic substances tested, only pyruvate stimulated nitrogen fixation. Stimulation was observed both in the light and the dark and was most pronounced under photosynthesizing conditions. This indicates that the chief role of pyruvate was other than to act as a carbon skeleton. More pyruvate was decarboxylated by nitrogen-starved cells, which also fixed nitrogen more vigorously than normal cells. It was concluded that pyruvate may supply cofactors for nitrogen reduction in *A. cylindrica*.

In the presence of pyruvate in the light, oxygen evolution was stimulated. This effect was interpreted as the reassimilation of carbon dioxide evolved from pyruvate by the phosphoroclastic reaction. In consequence, observations of oxygen evolution concomitant with nitrogen uptake, should not be considered definite evidence in favour of reports of photochemical nitrogen reduction.

REFERENCES

Allen, M. B. and Arnon, D. I. (1955). *Pl. Physiol.* **30**, 366.
Allison, F. E., Hoover, S. R. and Morris, H. J. (1937). *Bot. Gaz.* **98**, 433.
Arnon, D. I., Losada, M., Nozaki, M. and Tagawa, K. (1961). *Nature, Lond.* **190**, 601.
Bennett, R., Rigopoulos, N. and Fuller, R. C. (1964). *Proc. natn. Acad. Sci. U.S.A.* **58**, 762.
Burris, R. H. and Wang, L. C. (1960). *Pl. Physiol.* **35**, suppl. xi.
Carnahan, J. E., Mortenson, L. E., Mower, H. F. and Castle, J. E. (1960). *Biochim. biophys. Acta* **44**, 520.
Cobb, H. D. and Myers, J. (1964). *Am. J. Bot.* **51**, 753.
Cox. R. M. (1966). *Arch. Mikrobiol.* **53**, 263.
Cox, R. M. (1967). *Arch. Mikrobiol.* **56**, 193.
Cox, R. M. and Fay, (1967). *Arch. Mikrobiol.* **58**, 357.
Cox, R. M., Fay, P. and Fogg, G. E. (1964). *Biochim. biophys. Acta* **88**, 208.
Davis, E. B.and Tischer, R. G. (1966). *Nature, Lond.* **212**, 302.
Dilworth, M. J. D., Subramanian, D., Munson, T. O. and Burris, R. H. (1965). *Biochim. biophys. Acta* **99**, 486.
Fay, P. (1965). *J. gen. Microbiol.* **39**, 11.
Fay, P. and Cox, R. M. (1966). *Biochim. biophys. Acta* **126**, 402.
Fay, P. and Fogg, G. E. (1962). *Arch. Mikrobiol.* **42**, 310.
Fogg, G. E. (1961). *Bact. Rev.* **20**, 148.
Fogg, G. E. and Than-Tun (1958). *Biochim. biophys. Acta* **30**, 209.
Fogg, G. E. and Than-Tun (1960). *Proc. R. Soc. B153*, 111.
Hardy, R. W. F. and Knight, E. (1966). *Biochim. biophys. Acta* **132**, 520.
Kessler, E. (1964). *A. Rev. Pl. Physiol.* **15**, 57.
Mortenson, L. E. (1964). *Proc. natn. Acad. Sci. U.S.A.* **52**, 272.
Myers, J. (1960). *In* "Encyclopaedia of Plant Physiology", Vol. IV, Part I (Ruhland, W. ed.) p. 211, Springer-Verlag, Berlin.
Pardee, A. B. (1949). *J. biol. Chem.* **179**, 1085.
Pratt, D. C. and Frenkel, A. W. (1959). *J. biol. Chem.* **34**, 333.
Wolfe, M. (1954). *Ann. Bot.* **18**, 309.

Section 1d

c

Nitrite and Hydroxylamine in Inorganic Nitrogen Metabolism with Reference Principally to Higher Plants

E. J. HEWITT, D. P. HUCKLESBY AND G. F. BETTS

Long Ashton Research Station
University of Bristol, England

INTRODUCTION

Nitrite is the stable first intermediate in the assimilation of nitrate by plants, fungi and bacteria. Nitrite rarely occurs in tissues in high concentrations under normal conditions whereas nitrate accumulation to high concentrations (0·1 M) is quite often observed, particularly under conditions of molybdenum deficiency. Spencer and Wood (1955) found that if tomato seedlings, chlorotic because of molybdenum deficiency, were given the element in light, nitrite rapidly appeared and disappeared again in about 2 h. Ammonia accumulated steadily over several hours and nitrate steadily disappeared. Reduction of nitrite is more rapid than that of nitrate once the enzyme system begins to operate even when chlorophyll is deficient. Kessler (1953) first showed that nitrite accumulated in the dark and disappeared in the light in cultures of *Ankistrodesmus*. Vanecko and Varner (1955) showed that nitrite reduction by wheat leaves was dependent on light and resulted in oxygen evolution. Similar results were obtained by Kessler (1955) who found that nitrite served as an efficient Hill reagent for oxygen evolution by *Scenedesmus braunii* and that light immediately stimulated nitrite reduction. Huzisige and Satoh (1960) observed a similar light dependence of nitrite reduction by *Euglena gracilis*. Huzisige and Satoh (1961) then showed that nitrite was reduced to ammonia by spinach chloroplast grana when illuminated in the presence of an additional soluble protein fraction. C. F. Cresswell (Thesis 1961, University of Bristol) found that when marrow leaf discs were infiltrated with nitrite, this was rapidly lost during illumination but only slowly disappeared in darkness. In green plants and algae therefore physiological nitrite reduction is mainly dependent on a photochemical system unlike the mechanisms in fungi, yeasts and non-photosynthetic bacteria.

The intermediates between nitrite and ammonia have not been identified unequivocally in plants or fungi. Meyer and Schulze (1894) proposed that nitrite reduction proceeds by a sequence of three $2e^-$ steps *via* two intermediates, the first at the oxidation level of hyponitrite and the second being hydroxylamine. The occurrence of hydroxylamine in tissues of higher plants has been reported occasionally in earlier papers but Wood (1953) in a critical review concluded that free hydroxylamine which is highly toxic is unlikely to be present in appreciable concentrations. The presence of oximes was however regarded as likely and consistent with the supposed production of hydroxylamine as an intermediate in nitrite reduction by green plants as well as by micro-organisms.

Virtanen (1961) held the same opinion regarding the significance of oximes and hydroxamates produced by *Torulopsis utilis* and rye roots grown with nitrate or nitrite especially as these products were not detected when only ammonium nitrogen was given. The possibility was considered that normally hydroxylamine remains bound to the reductive enzymes and does not appear as a free intermediate.

Nason *et al.* (1954) reported that extracts of soya bean leaves and *Neurospora crassa* felts contained enzyme systems that reduced nitrite and hydroxylamine to ammonia, though no data were given regarding hydroxylamine. NAD(P)H and FMN (or FAD) produced some increase in ammonia formation from nitrite over high endogenous rates and Mn^{2+} stimulated, while cyanide inhibited the systems. Hydroxylamine was said to be produced (as detected by the Csáky oxidation by iodine) from nitrite, and Mn^{2+} was said (without presenting data) to be stimulatory in hydroxylamine reduction to ammonia. On the basis of these and comparable statements for *N. crassa* Nason *et al.* (1954) postulated that nitrite reduction to ammonia probably proceeds by a sequence of $2e^-$ steps *via* an unknown intermediate and hydroxylamine. Nevertheless later work by Roussos and Nason (1960) failed to detect any ammonia formation from nitrite or hydroxylamine in the presence of soya bean protein, Mn^{2+} and an unidentified cofactor which together caused loss of NADH in presence of nitrite or hydroxylamine. Nitrite did not suffer any net loss, and though hydroxylamine disappeared no product was identified. Hageman, Cresswell and Hewitt (1962) suggested that the Mn-dependent peroxidation of hydroxylamine described by Cresswell and Hewitt (1960) might be concerned in these changes.

In comparable work with *N. crassa*, Medina and Nicholas (1957a) reported enzymic reduction of hyponitrite to ammonia (0·1 μmole/h/mg protein) in the presence of NADH and FMN (FAD). Hydroxylamine was detected by Fe^{3+} and *o*-phenanthroline but the amount produced was not specified. Medina and Nicholas (1957b) measured rates of ammonia formation from nitrite, hyponitrite and hydroxylamine with *N. crassa* preparations when no fractionation of possibly sequentially related enzymes had been attempted. Hyponitrite (at $6{\cdot}7 \times 10^{-4}$ M) yielded ammonia at about 60% of the rate observed with nitrite (at $5{\cdot}7 \times 10^{-4}$ M) on a specific activity basis. They stated that the reduction product of nitrite inhibited nitrite reduction and concluded that this was hyponitrite. Nicholas (1959) showed that hyponitrite caused 60% inhibition of nitrite reductase when present at a 2·5 molar excess ratio to nitrite whereas hydroxylamine at forty times nitrite was not inhibitory. Purified nitrite reductase was also said not to be able to reduce hyponitrite. Nicholas *et al.* (1960) concluded similarly that a decline in rate of nitrite reduction after 15 min indicated the accumulation of an inhibitory product. Hyponitrite (5×10^{-4} M) was inhibitory. The K_m for nitrite was $1{\cdot}1 \times 10^{-4}$ M but the nitrite concentration used for these tests was shown as 0·02 μmoles in 1 ml, i.e. 2×10^{-5} M.

Zucker and Nason (1955) first described a hydroxylamine reductase from *N. crassa* and the presence of this was confirmed by Medina and Nicholas (1957b) who found that reduction of hydroxylamine (at $5{\cdot}7 \times 10^{-2}$ M) to ammonia was four times as fast as that of nitrite (at $5{\cdot}7 \times 10^{-4}$ M). The K_m values for hydroxylamine for at least two enzymes found in *N. crassa* have now been determined as $1{\cdot}4 \times 10^{-2}$ M and $9{\cdot}6 \times 10^{-4}$ M by Leinweber *et al.* (1965) and Siegel *et al.* (1965). The value of $3{\cdot}8 \times 10^{-3}$ M reported by Zucker and Nason (1955) was therefore probably a complex value but the lower K_m value is still nine times greater than for nitrite ($1{\cdot}1 \times 10^{-4}$ M) found by Nicholas *et al.* (1960). For hydroxylamine to be an intermediate in the reduction of nitrite to ammonia by the system of Medina and Nicholas (1957b), hydroxylamine would have to be reduced at about 12% of $V_{max.}$ (assuming $5{\cdot}7 \times 10^{-2}$ M to be near saturation) when present at an actual concentration not exceeding, and almost certainly very much less than, 5% of the K_m value.

Nicholas (1957, 1958) inferred, from the results of the experiments with his associates reviewed above, and then concluded explicitly (Nicholas 1959) that a series of enzymes in plants—(not qualified by any phylogenetic distinction)—mediate the reduction of nitrite to ammonia *via* hyponitrite and hydroxylamine as the physiologically important route.

The significance of hyponitrite in nitrite metabolism of higher plants was rejected by Frear and Burrell (1958). They pointed out its great instability at physiological pH values and concluded that when introduced into plant leaves it was first oxidized to nitrite before reappearing as ^{15}N-labelled ammonia. The instability of hyponitrite at pH 7–7·5 was further shown by Anderson (1963, 1964).

Vaidyanathan and Street (1959) found that aqueous extracts of sterile cultured excised tomato roots catalyzed the disappearance of nitrate, nitrite, hyponitrite and hydroxylamine in the presence of NADH, nicotinamide, FMN, 2×10^{-5} M Mn^{2+} and supplementary electron donors. Loss of nitrite exceeded 1·5 μmole/h/g fresh wt. It was Mn-dependent and the main product yielded nitrite after re-oxidation by iodine in acetic acid. Hydroxylamine could not be detected with Fe^{3+} and *o*-phenanthroline and only 2% (0·03 μmoles) was recovered as ammonia. Hyponitrite (1 μmole in 5 ml) disappeared very slowly and yielded only 2% (0·02 μmoles) of ammonia in 10 h. Hydroxylamine (4·3 μmoles) disappeared in 2 h but yielded, though more, still only 30–35% of the expected quantity of ammonia, and oxime formation was suggested. They concluded from this work that in tomato roots, nitrite reduction is Mn-dependent and that hyponitrite reduction is the rate-limiting step.

Fewson and Nicholas (1960) reported that homogenates of roots and leaves of plants contained an NADH-enzyme which reduced nitric oxide measured as gas uptake at rates between 0·3 to 1·5 μmoles/h/mg nitrogen present in the homogenate but the product was not determined.

The principal conclusion of Nason *et al.* (1954) that nitrite reductase of plants was an NAD(P)H, flavin and Mn^{2+}-dependent system has not been confirmed by later investigations. Hageman *et al.* (1962), Cresswell *et al.* (1962), Cresswell *et al.* (1965) showed that the free radical of reduced benzyl viologen (BV·) reduced nitrite to ammonia quantitatively in an enzymic system. Nason (1962) first suggested that the physiological electron carrier in photochemical nitrite reduction which was probably operating in the system described by Huzisige and Satoh (1961) might well be the factor of San Pietro and Lang (1958) then known as "photosynthetic pyridine nucleotide reductase" (PPNR). The obvious similarities between PPNR and certain iron-proteins of bacteria now classified as ferredoxins (Tagawa and Arnon 1962) and the methyl and benzyl viologens (MV, BV) in both dark and chloroplast reactions led to the independent demonstration by Hewitt and Betts (1963), Huzisige *et al.* (1963), Losada *et al.* (1963), Paneque *et al.* (1963) and Paneque *et al.* (1964), that ferredoxin is the physiological carrier in both photochemical and dark (NADPH) systems for reduction of nitrite to ammonia in higher plants. The purification and some properties of nitrite reductase, principally from spinach, have been described by Losada *et al.* (1965), Betts and Hewitt (1966), Joy and Hageman (1966), Ramirez *et al.* (1966) and Shin and Oda (1966). Comparative studies on nitrite and hydroxylamine reduction in photochemical or dark systems were made by Betts and Hewitt (1966), Hewitt and Hucklesby (1966), Hewitt *et al.* (1966), with respect to protein fractionation, ferredoxin requirements, other electron carriers and rates of ammonia formation. Details of some further experiments are presented here.

METHODS

PLANT MATERIAL

Spinach (*Spinacea oleracea* Summer Variety) and marrow (*Cucurbita pepo*

Sutton's Improved Green Bush) were grown in sand culture with nitrate as nitrogen source (Hewitt, 1966). Rapidly expanding leaves were used for enzyme preparations and mature leaves or commercial spinach were used for ferredoxin extraction.

ANCILLARY PROTEINS

Ferredoxin was prepared from frozen stored leaves, Davenport (1960), from both marrow and spinach by a composite procedure of Betts and Hewitt (1966) based on methods of several workers and further modified here by using Sephadex G25 for removal of acetone and pigments, for desalting after elution from the initial DEAE cellulose columns and after extraction of the protamine precipitate (San Pietro and Lang, 1958). The products after final chromatography on DEAE cellulose and dialysis had the following spectrophotometric absorption ratios and were relatively pure by the criteria of Tagawa and Arnon (1962), San Pietro (1963) and Fry and San Pietro (1963).

nm	260	278	330	420	460
Spinach	0·80	1·0	0·65	0·46	0·42
Marrow	0·82	1·0	0·64	0·46	0·41

Plastocyanin was obtained in high yields from marrow as a by-product of ferredoxin preparations. It was purified by ammonium sulphate precipitation (60% to saturated) and chromatography on DEAE cellulose with phosphate buffers (Katoh *et al.*, 1962). Excess potassium ferricyanide was added at all stages until final dialysis. Repeatedly purified preparations had a characteristic major absorption band at 597 nm, 67% of that at 278 nm, and showed fine structure absorption bands at 253, 259, 266 and 285 nm. The subsidiary bands at 462 and 775 nm were relatively weak.

NADP-ferredoxin oxido-reductase (diaphorase) was also a by-product of ferredoxin preparations from spinach and marrow. It was purified by further precipitation by ammonium sulphate (40–60%), adsorption on DEAE cellulose, washing with 0·04 M and elution by 0·08 M NaCl (Shin *et al.*, 1963), and re-chromatography and elution by 0·14 M NaCl. This preparation was then free of nitrite and hydroxylamine reductase activity. The specific activity was about 1 μmole FMN reduced/min/mg at 20°C and characteristic flavoprotein bands were present at 455 and 385 nm.

CHLOROPLAST GRANA

Spinach leaves (100/150 g) were ground at 0°C in a large chilled mortar in 4 vol. 0·4 M sucrose, 0·05 M Tris HCl pH 7·8, 0·01 M NaCl (Davenport, 1960).

Chloroplasts were disrupted in 0·003 M Tris HCl, and grana were resuspended twice in the same medium. They were finally resuspended in 0·07 M Tris buffer i.e. swollen (Spencer and Unt, 1965). The suspension was heated for 5 min at 55°C to destroy oxygen evolution capacity (Paneque and Arnon, 1962) and to inactivate residual endogenous nitrite reductase (Betts and Hewitt, 1966). Chlorophyll was determined according to Arnon (1949). Spinach grana, when not used for oxygen evolving reactions, were stable for several hours at 4°C (Vernon and Zaugg, 1960), and were used for routine assays with spinach and marrow ferredoxins and enzymes. Unheated grana were only used for specific tests described later. Marrow chloroplast grana used in some initial experiments were re-suspended in 0·05 M Tris and frozen in batches in acetone-solid CO_2 while spinning centrifugally (Clendenning and Gorham, 1950; Duane and Krogmann, 1963) to preserve activity.

ENZYME ASSAYS

Several assay systems were used according to circumstances. Benzyl viologen was useful for monitoring of column effluent fractions and checking for activity in routine purifications before using physiological systems. Ferredoxin, reduced by a dark system with glucose-6-phosphate or photochemically, was used for determining specific activity during fractionation, and FMN was used for certain assays as described later. The unit of specific activity is 1 μmole nitrite, or hydroxylamine reduced /min/mg protein at the specified temperature.

Benzyl viologen. Reduced benzyl viologen (BV·) was prepared as described by Hageman *et al.* (1962), Hewitt and Nicholas (1964). It was used as electron donor under hydrogen at substrate concentrations either as in the work of Cresswell *et al.* (1965) in 7 ml or in a rapid spectrophotometric monitoring method in 3 ml in 12 mm diameter Thunberg tubes with an adapted EEL "Spectra". The validity of BV· for the assay of nitrite reductase during fractionation was demonstrated by Betts and Hewitt (1966).

Glucose-6-phosphate systems. Ferredoxin (usually 2×10^{-5} M) from the relevant species, or FMN (5×10^{-4} M) were used as electron donors under argon in the presence of glucose-6-phosphate (2×10^{-2} M) and NADP ($2{\cdot}5 \times 10^{-4}$ M) in 2 ml together with glucose-6-phosphate dehydrogenase (BDH) and diaphorase at a 4- to 5-fold excess over the maximum rate of electron transport for nitrite reduction to ammonia. The dehydrogenase was dialyzed to remove ammonia when this was to be determined. The electron donor complex was incubated 10 min at 25°C before tipping the substrate and incubating for a further period of 10–25 min. Substrate loss was linear with time and showed no lag period. Activity was proportional to protein but the relationship was perceptibly curvilinear in spite of the considerable excess of electron donating capacity.

Photochemical systems. Spinach grana (heated 5 min at 55°C) equivalent to 100 to 1000 μg chlorophyll in 2 ml 0·05 M Tris HCl pH 7·8, 10^{-2} M ascorbate (made up as 0·1 M at pH 7·8 in phosphate buffer) and 7×10^{-5} M dichlorophenol-indophenol were electron donors for chlorophyll A (Vernon and Zaugg, 1960; Losada *et al.*, 1961), together with usually 10^{-5} M plastocyanin and between 2×10^{-5} and 10^{-4} M ferredoxin of the relevant species. Nitrite was present at 7×10^{-4} and hydroxylamine at 10^{-3} M. Assays were carried out at 20°C under argon unless otherwise stated for times between 1 and 10 min as specified. Illumination was by a 250 W tungsten projector lamp providing up to 90,000 lux measured by a miniature photoelectric cell mounted in a blank assay tube and calibrated with an EIL "Light Master". The reaction was started by tipping substrate immediately before illumination. The reaction was stopped by excluding light and determining both nitrite and hydroxylamine by the alkaline iodine procedure (Hewitt and Betts, 1963), which avoids interference by ascorbic acid.

Loss of substrate was measured against dark time and no-enzyme controls. The difference between these for nine tests was $14{\cdot}0 \pm 9{\cdot}6$ nmoles/min with 10^{-4} M ferredoxin, 1 mg chlorophyll and 90,000 lux and was 3% of the most active activities. In three of these tests and at several other ferredoxin concentrations the loss with no enzyme was nil and it therefore seems unlikely that ferredoxin, highly reduced in a photochemical system, reacts directly with nitrite, contrary to the conclusions of Joy and Hageman (1966) that some reaction occurred when ferredoxin was reduced by dithionite. The low activities observed in the photochemical controls with no enzyme can be attributed to traces of endogenous enzymes remaining in the washed grana where the activity was shown to be located by Betts and Hewitt (1966), Ramirez *et al.* (1966) and Ritenour *et al.* (1967). Activity was proportional to protein up to amounts producing a loss of 500 nmoles nitrite/min/mg chlorophyll, with 90,000 lux and certain pH values (7·6–7·9) as indicated in the Results but the system was easily overloaded by enzyme as found also by Huzisige and Satoh (1961) and Shin and Oda (1966).

ELECTROPHORESIS

Electrophoresis was done on 0·2 ml samples fortified by 5% sucrose (Reisfield, Lewis and Williams, 1962), using a Shandon disc electrophoresis apparatus. Acrylamide gel columns, $0{\cdot}5 \times 5{\cdot}0$ cm containing 7% cross linkage, were used at 5 m.a. and pH 8·3 in Tris-glycine buffer (Ornstein and Davis, 1964), and protein was stained by Naphthalene Black 12B (BDH).

COLUMN CHROMATOGRAPHY

DEAE cellulose powder (Whatman DE11) was purified before use by prolonged stirring in N KOH, quick suspension in 0·1 N HCl and re-suspension in 0·1 N KOH before washing with copious amounts of water and titration to pH 7·5 in 0·005 M Tris HCl buffer. The molecular exclusion chromatography for molecular weight fractionation was carried out on columns of Sephadex G100 or G200 using the procedures of Andrews (1964, 1965). Gels were carefully equilibrated and swollen for several weeks at 1 °C before packing the columns. These were repeatedly calibrated with seven or more proteins ranging in molecular weight from human γ-globulin to mammalian cytochrome c. Samples (7 to 18 ml) were run in 0·03 M $KHPO_4$ pH 7·5,0·1M KCl at 20 ml/h. Fractions (3 ml) were subdivided after collection and frozen until required for assay by comparative methods.

OTHER DETERMINATIONS AND DETAILS

Ammonia production was measured in Conway microdiffusion cells (Conway, 1962; Hewitt and Nicholas, 1964). Protein was determined by the Folin phenol method of Lowry *et al.* (1951) and by the spectrophotometric methods of Warburg and Christian (1941) and Patterson (1964) with Unicam recording SP 700 and SP 800 instruments when limited samples of highly purified preparations were required for several different purposes. Nitrite and hydroxylamine were determined as described by Cresswell *et al.* (1965) except when the photochemical system with ascorbate was used as noted above. Assay mixtures were diluted afterwards by a convenient volume of water (2–5 ml) to obtain suitable concentrations for the determinations. Enzyme preparations were relatively stable at −15°C but repeated thawing and freezing or storage at 0°C for prolonged periods were avoided as far as possible by freezing small subsamples and drawing upon these just before use.

RESULTS

PROTEIN FRACTIONATION

Several methods of fractionation were used with three objectives: to increase specific activity; to study relationships between nitrite and hydroxylamine reducing activities; to study electron donor requirements.

Purification procedures. Tables I and II. Spinach leaves (200–400 g) were macerated in 2 wt. vol. of 0·1 M Tris HCl pH 7·8 at 1°C, strained and centrifuged. Protein was precipitated by 48–70 % saturated ammonium sulphate or at −10°C by 33–60 % acetone. Protein was dissolved in a minimum volume

TABLE I

Fractionation and activity of nitrite reductase from spinach assayed by photochemical reduction of ferredoxin.

Systems contained 800 μg chlorophyll as washed heated grana, 8×10^{-5} M ferredoxin, 6×10^{-6} M plastocyanin, 10^{-3} M ascorbate. 7×10^{-5} M dichlorophenol-indophenol, 0·05 M Tris-HCl pH 7·7, 0·01 to 1·6 mg enzyme protein in 2·0 ml. Illumination was for 3 min at 20°C with 85,000 lux under argon.

Fraction		Volume (ml)	Protein (mg/ml)	Units	Specific activity (μmoles/min/mg)	Recovery %
Initial extract		900–1500	7·1–17·2	300–1600	0·059 ±0·019 (6)	—
48–70% ammonium sulphate ppt. desalted by Sephadex G25		110–240	4·5–15·6	41–370	0·16 ±0·04 (7)	14–23
Adsorption, washing and elution on DEAE cellulose		8–140	0·3–3·7	14–96	1·42 ±0·78 (12)	1·5–6·0
Re-chromatography of bulked samples of preceding stage	(a)	20	0·14–0·18	15–17	6·0 ±0·50 (4)	3·0
	(b)	10	0·21	21	10·0 — (1)	
Chromatography on Sephadex G100 of samples of preceding stage; peak fractions	(a)	5	0·07	8·4	23·9 — (1)	1·5
	(b)	5	0·08	7·8	19·5 — (1)	
Concentration of G100 Sephadex fractions around peaks, by adsorption and elution on DEAE cellulose		3·6	0·034	6·2	57·0 — (1)	0·6

TABLE II

Comparative fractionation and activities of nitrite and hydroxylamine reductases from spinach assayed by ferredoxin in presence of glucose-6-phosphate and NADP.

Systems contained 2×10^{-5} M ferredoxin, 2×10^{-2} M glucose-6-phosphate, $2{\cdot}5 \times 10^{-4}$ M NADP, 5-fold excess glucose-6-phosphate dehydrogenase and diaphorase, 0·05 M phosphate pH 7·5 and 0·03 to 1·3 mg protein. Assays were for 10 min at 25°C after pre-incubation to reduce electron donors as described in Methods.

Fraction	Protein mg/ml	(a) Nitrite Reductase Units	(a) Nitrite Reductase Spec. activity	(b) Hydroxylamine Reductase Units	(b) Hydroxylamine Reductase Spec. activity	Ratio a/b
Initial extract	6·0–13·3	50–280	0·0095 ±0·0016 (5)	40–850	0·035 ±0·030 (4)	0·37
48–70% ammonium sulphate ppt. desalted by Sephadex G25	9·5	43	0·064 —	72	0·11	0·58
33–60% acetone ppt. alternative to above	11·1–12·5	15–120	0·063 ±0·005 (3)	—	—	—
Acetone ppt. desalted by Sephadex G25	8·3–10·0	21–133	0·071 ±0·014 (4)	11–30	0·021 ±0·006 (3)	3·4
Adsorption, washing and elution on DEAE cellulose	0·1–1·6	1–8	0·63 ±0·22 (4)	0·3–2	0·13 ±0·033 (3)	4·8
Re-chromatography of bulked samples of preceding stage	(i) 0·48 (ii) 0·15	0·5 1·0	0·96 — 1·70 —	0·4	0·76 —	2·2
Chromatography on Sephadex G100 of sample of preceding stage	0·066	0·67	2·24 —	0·025	0·085 —	26·5
Concentration of active fractions from above on DEAE cellulose	0·034	0·72	6·0	0·12	1·0 —	6·0

of 0·01 M Tris or phosphate buffer pH 7·5, desalted and decolourized by Sephadex G25 and adsorbed on a DEAE cellulose column (2·5 × 10 cm). Ammonium sulphate was preferred because acetone often caused interference in hydroxylamine reductase assays which was not readily eliminated by Sephadex or dialysis treatments. For maximal purification acetone was preferable with spinach but was unreliable with marrow probably because of phenolic-protein associations.

Protein on DEAE cellulose was washed with 5 vol 0·12 M NaCl in 0·007 M Tris HCl which removed the NADP-ferredoxin oxido-reductase and caused losses of hydroxylamine reductase. Nitrite reductase was eluted by 0·2–0·25 M NaCl (in Tris buffer pH 7·7) and collected in 10–15 ml fractions. Best fractions were bulked, desalted on Sephadex G25 and re-chromatographed on DEAE cellulose, by washing with 0·16 M and eluting with 0·2 M NaCl.

Selected fractions (7–15 ml) were passed through a column of Sephadex G100 (230 ml) prepared and operated as specified by Andrews (1964, 1965) and described in the Methods section. Peak fractions of nitrite reductase were identified by the BV· spectrophotometric assay and sampled for assay by ferredoxin in the glucose-6-phosphate and photochemical systems. The remaining adjacent fractions were then bulked (60 ml), readsorbed on DEAE cellulose (5 × 0·5 cm) and washed with 50 ml 0·13 M NaCl. The column was extruded and centrifuged to expel surplus fluid. The cellulose was then extracted three times by 0·25 M NaCl. The extract (3·5 ml), filtered through Whatman 541 paper, and a control specimen were used for spectrophotometric, electrophoretic and activity tests.

Purification of marrow enzymes (to be described fully elsewhere) was basically similar but comprised use of Darco G60 charcoal to decolourize initial extracts; Sephadex G100 for combined desalting and protein fractionation after ammonium sulphate; adsorption and elution from hydroxylapatite followed by adsorption and elution from DEAE cellulose.

The maximum specific activities for nitrite reductase so far obtained by the use of photochemical assay are 52–57 for spinach, Table I, and 48 for marrow; as μmoles reduced/min/mg protein at 20°C. These values represent overall increases of 970 for spinach and 760 for marrow. Using the glucose-6-phosphate system the maximum specific activities at 25°C are 6·0 for spinach (Table II) and 3·3 for marrow, representing increases of 620 and 1250 respectively. An increase of 540 for spinach to a value of 4·1 was recorded by Ramirez *et al.* (1966) using dithionite and methyl viologen under aerobic conditions at 30°C. Joy and Hageman (1966) obtained an increase of 90 for maize to a value of 0·35 using glucose-6-phosphate and ferredoxin *plus* BV· at 25°C. Shin and Oda (1966) recorded a value for spinach of 1·0 for a photochemical assay at 20°C.

TABLE III

Recoveries of ammonia from nitrite or hydroxylamine using preparations from spinach and marrow which were highly purified with respect to nitrite reductase activity.

Recoveries were measured in two assay systems: A, ferredoxin reduced by glucose-6-phosphate and NADP; B, ferredoxin reduced photochemically as described in the Methods. Initial substrate levels: nitrite 1500; hydroxylamine 2000 nmoles. Purification factors refer to increased specific activity of nitrite reductase.

Assay method	Source	Purification factor	Substrate	Loss nmoles	Time (min)	Protein (mg)	Ammonia gain nmoles	Ammonia gain %
A	Spinach	250	nitrite	1493 ± 8 (3)	11–20	0·066	1477 ± 6 (3)	99
			hydroxylamine	120 ± 40 (3)	20	0·066	115 ± 39 (3)	95
A	Spinach	620	nitrite	413	25	0·0034	392	95
			hydroxylamine	67	25	0·0034	78	116
B	Spinach	530	nitrite	1030	3	0·011	972	94
			nitrite	891	2	0·015	854	96
			hydroxylamine	12	3	0·015	10	83
B	Spinach*	970	nitrite	714	5	0·003	675	95
A	Marrow	1250	nitrite	1232	25	0·015	1171	95
			hydroxylamine	212	25	0·015	380	185
B	Marrow	720	nitrite	416	3	0·0030	450	108
		760	nitrite	295	3	0·0028	312	106
		720	hydroxylamine	11	3	0·0030	33	330

* This preparation was thawed and held at room temperature for 2 h, refrozen and thawed before using for this experiment; this treatment caused some loss in activity compared with the value of 57 for specific activity reported in the text for the original assay.

Purification of hydroxylamine reductase from spinach was less efficient and less consistent that that of nitrite reductase. As seen in Table II and as shown by Betts and Hewitt (1966) there was a progressive and large change in the ratio of the two activities during fractionation. Table III shows that hydroxylamine activity could be almost totally eliminated from nitrite reductase of high specific activity. This effect was reported also by Ramirez *et al.* (1966). The maximum purification of hydroxylamine reductase from spinach was about 29 in our present work when assayed with glucose-6-phosphate and 18 in that of Betts and Hewitt (1966) with photochemical reduction of ferredoxin. Results with marrow however have shown a 500-fold purification of hydroxylamine reductase assayed by ferredoxin and the glucose-6-phosphate system. Further points regarding the relative activities of these enzymes are noted in connection with molecular exclusion chromatography.

MOLECULAR EXCLUSION CHROMATOGRAPHY AND ELECTRON DONOR SPECIFICITY

Relationships between activity and protein fractions. Betts and Hewitt (1966) reported a separation of nitrite and hydroxylamine reductase activities in spinach preparations by chromatography on Sephadex G200, nitrite reductase being the less retarded. There were also substantial increases in specific activities and this aspect has been exploited in the results of the previous section. Further work by Hewitt and Hucklesby (1966) using G100 and G200 showed that nitrite reductase (NR) activity of spinach (Fig. 1) and marrow was practically confined to single symmetrical peaks. These correspond to molecular weights of about 64,000 for spinach and 60,000 for marrow when either BV· or ferredoxin (reduced either photochemically or by glucose-6-phosphate and NADP) were used. No other peaks were found down to 10,000 mol. wt. using BV· or down to 27,000 (the lowest assayed) using ferredoxin.

Hydroxylamine reductase activity occurred in two major peaks reacting with BV· These corresponded to mol. wt. of about 64,000 (or sometimes slightly less) (HR1) and about 32,000 (HR2) for spinach and to mol. wt. of between 50,000 and 60,000 (HR1) and about 32,000 (HR2) for marrow. In spinach the 64,000 peak was the less abundant of the two; the ratio HR1 : HR2 varying between 0·1 and 0·5; HR1 was initially overlooked in the preparation used by Betts and Hewitt (1966). In marrow HR1 was more abundant, the ratio of HR1 : HR2 varying between 0·5 and 1·4. The greater discrepancy between independent values with BV· assays for HR1 peaks from marrow may reflect effects of two separate factors; namely continued improvements in precision of the fraction collector siphon and

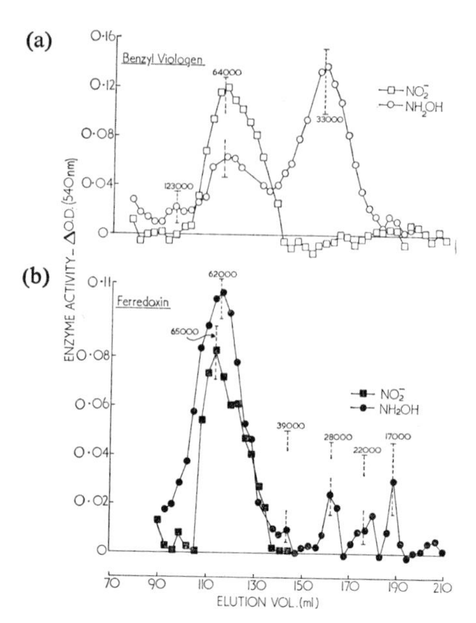

FIG. 1. Chromatography of nitrite and hydroxylamine reductases from marrow and spinach on Sephadex G100 and G200. (a) assayed by BV· in subdivided fractions. (b) assayed by ferredoxin on similar divided fractions from a duplicate series.

calibrations, and variable proportions of HR1 and HR2. There was in addition a possible lesser peak at about 100,000 to 123,000 mol. wt. in both species. There were often appreciable separations between NR and HR1 in marrow when assayed by BV· which could have resulted from displacement of the HR1 peak to a lower mol. wt. value by the mutual supplementation effect of the overlapping parts of HR1 and HR2. When ferredoxin was used no separation within experimental limits, even using a column of 570 ml, was observed between HR1 and NR from either plant.

Electron donor specificity. Nitrite reductase was found to be unreactive with FMN or FAD in the work of Cresswell *et al.* (1965) Joy and Hageman (1966) Ramirez *et al.* (1966). In our later work we have found that FMN or FAD produced between 5% and 30% of the activity obtained with BV· or ferredoxin. For either species the effect of flavins was unpredictable and may indicate the variable presence of an unknown intermediate carrier. For this reason no assays with FMN have been carried out with the NR fractions. In contrast with NR it was found that FMN was an efficient electron carrier for the HR1 fraction of hydroxylamine reductase (Fig. 2); there was however practically no activity in the HR2 region although several minor peaks appeared to be reactive with FMN. This result was

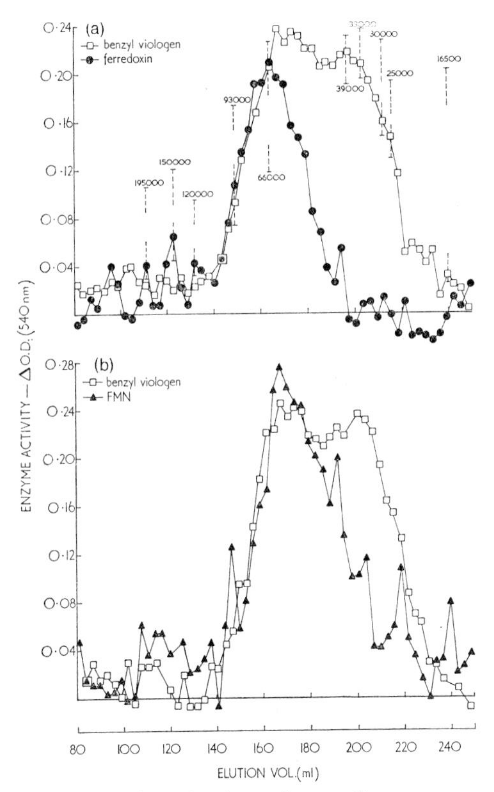

FIG. 2. Chromatography of hydroxylamine reductase from marrow on Sephadix G200. (a) assayed by BV· and ferredoxin in sub-divided fractions. (b) assayed by BV· and FMN on similar divided fractions from a duplicate series.

obtained with preparations from both plants. The distribution of the FMN reactive peaks suggested approximate multiples of about 16,000 mol. wt. intervals in marrow and a possibly more complex distribution in spinach, in each case suggestive of an oligomeric series.

When ferredoxin was used the HR2 peak was even more sharply resolved and there was less activity in the lower mol. wt. regions than with FMN. There were however other lesser peaks for both plants between 40,000 and 10,000 and those at 40,000 and 22,000 mol. wt. regions were most consistent for spinach. The results of eight experiments are summarized in Fig. 3 where the evidence for an oligomeric series of HR fractions corresponding to units of 1, 2, 4 and 8 mol. wt. intervals is apparent as a minimum representation of a possibly more complex system.

The results described above explain the observation of Betts and Hewitt (1966) that whereas BV· and ferredoxin maintained a constant activity

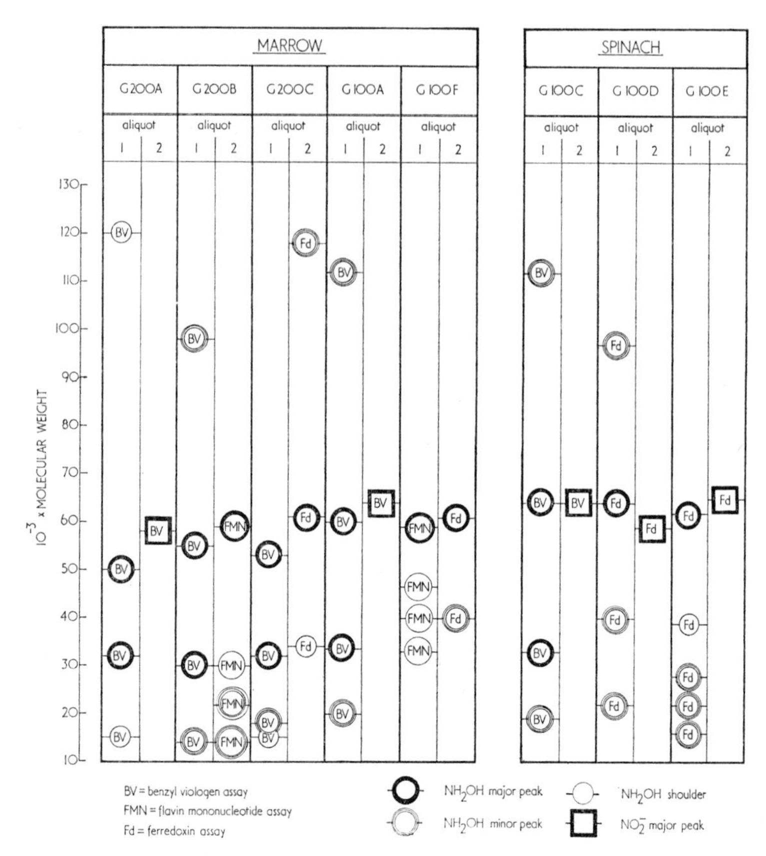

FIG. 3. Results of eight chromatographic experiments with G200 or G100 Sephadex columns with marrow or spinach preparations assayed for nitrite or hydroxylamine reductase activities by BV·, FMN or ferredoxin. The two sides of each column in the diagram represent assays on subdivided fractions. The centres of the symbols correspond with the estimates of molecular weight at the peak values on each occasion. Heavily outlined symbols represent major peaks; triple symbols represent clear minor peaks and single symbols represent shoulders or otherwise less clearly recognizable peaks. □, nitrite reductase; ○, hydroxylamine reductase.

ratio for nitrite reductase over a 55-fold purification, the ratio for the two donors changed considerably for hydroxylamine reductase when assayed simultaneously and consistantly increased in favour of ferredoxin. The change was pronounced after DEAE cellulose chromatography. The "leakage" of hydroxylamine reductase at low salt concentrations before elution of nitrite reductase and hydroxylamine reductase as a coincident peak when assayed by BV· is also explained by the same results above.

Although crude nitrite reductase (NR) showed variable weak activity with FMN and FAD, and HR1 reacted readily, attempts to remove a possible flavin prosthetic group by acid amonium sulphate precipitation of preparations containing both activities from both plants were unsuccessful. There was no restoration of denatured activity by adding flavins and no inhibition of either activity by several flavin analogues. Absorption spectra of most purified preparations obtained by us, or by Ramirez *et al.* (1966), show no indication whatever of flavoprotein peaks. The earlier report by Huzisige *et al.* (1963) that a flavoprotein was present in the reductase system can be attributed to the presence of the diaphorase. This is removed, as described in the Methods, by rather critical NaCl molarity ranges not used by Huzisige *et al.* and has been shown to be a separate protein also by Ramirez *et al.* (1966) and Joy and Hageman (1966).

The lower mol. wt. fraction HR2 does not react with ascorbate, cytochrome c, plastocyanin, α-lipoic acid, menadione or other vitamin K derivatives, NADH, or NADPH as electron donors, but dithionite will function in place of BV·. Preparations of high specific activity have a light absorption maximum at 399 nm (0·4 of that at 280 nm) which shifts to 415–430 nm on reduction. This could be a pyridoxal component with which hydroxylamine would probably react, and a pyridoxal oxime was suggested by Silver and McElroy (1954) to be involved in hydroxylamine reduction by *Neurospora crassa*. HR1 fractions do not however show this peak and its significance cannot yet be assessed.

PRODUCT OF REDUCTION

Ammonia was produced in quantitative yields from nitrite by preparations of high specific activity when ferredoxin was the electron donor reduced under physiological conditions by either glucose-6-phosphate and NADP or photochemically by chloroplast grana as shown in Table III and by Betts and Hewitt (1966) .The rate of hydroxylamine reduction was low or negligible compared with that of nitrite. There is at present still no evidence for the production of any "lesion" in the transfer of six electrons even in preparations purified over 1000-fold using a physiological reduction system and devoid of the dithionite or viologen carriers used by other workers. The conversion of nitrite to ammonia is independent of the rate of reduction when this is determined by the system used for reducing ferredoxin, i.e. by the NADP or photochemical procedures.

ELECTROPHORESIS

Preparations from both plants were examined by electrophoresis in acrylamide gel. Spinach preparations persistently showed two closely associated bands of relatively high mobility (R_f 0·65 to bromo-phenol blue) which together with a much fainter slower band, were the only ones visible at a concentration of 34 μg/ml after purification 870-fold (Table I). Marrow preparations of comparable purity (700–1200-fold) showed four bands of differing staining depth and more separated; a stronger one R_f 0·49 and weak ones R_f 0·32, 0·41 and 0·61. A purified preparation of HR2 from marrow showed a very strong band R_f 0·44, weaker bands at 0·48 and 0·65 and other very faint bands.

TURNOVER NUMBER

Minimum turnover numbers for nitrite reductase were based on the probable molecular weights of 64,000 for spinach and 60,000 for marrow and the maximum specific activities in the photochemical system. Values of $3{\cdot}5 \times 10^3$ and $2{\cdot}95 \times 10^3$ moles nitrite reduced/min/mole enzyme at 20°C were derived from spinach and marrow respectively. These represent values around 2×10^4 for electron transport to produce ammonia. In terms of photochemical activity rates up to 30 μmoles/h/mg chlorophyll equivalent to reduction of 90 μmoles/h of NADP were recorded.

FACTORS AFFECTING ACTIVITY

Effect of pH and protein concentration. The optimum pH for nitrite reduction in a spinach photochemical system varied with the amount of enzyme protein present as shown in Fig. 4. As enzyme was increased the pH

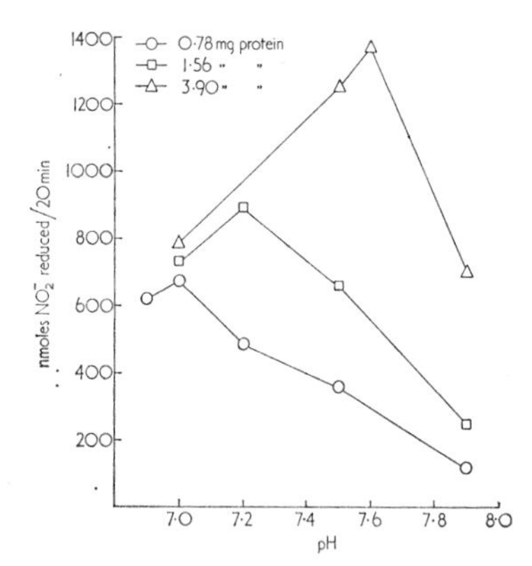

FIG. 4. Relationship between pH and protein concentration for nitrite reductase of spinach assayed by the photochemical system.

optimum shifted from about 7·0 to 7·7. This was interpreted as meaning that the pH optima for the photochemical reduction of ferredoxin and that for nitrite reductase were different and the net value depended on the balance between the relative capacities of the two components. As enzyme supply increased the effect of pH on the photochemical activity became dominant. Between pH values of 7·6 and 7·9 there was close proportionality between reductase activity and amount of protein over a five-fold range. The pH optimum for enzymic function of a marrow nitrite reductase was close to 7·0 when BV· was used. This value is complicated by the fact that BV· reacts directly with increasing rate with nitrite as pH values decrease below 7·0, and the enzymic values required to be corrected for this effect. Marrow hydroxylamine reductase showed possibly two optima with BV· at pH 6·7 and apparently at 8·7 through a range of buffer pH values produced by acetate, phosphate and borate.

Effect of temperature. In six pairs of comparisons at 20°C and 30°C with spinach preparations of diverse specific activity tested with 90,000 lux, 6×10^{-5} M ferredoxin, 3×10^{-6} M plastocyanin and 800–1000 μg chlorophyll as grana the value of Q_{10} was $2{\cdot}45 \pm 0{\cdot}20$. As the photochemical activity is not temperature dependent the quotient applies either to the reductase or to an effect of diffusion in the highly particulate system. If this factor is discounted an apparent activation energy of 15 kcal would be inferred.

Plastocyanin. The addition of plastocyanin between 3×10^{-6} and 10^{-5} M increased the overall activity of the standard photochemical system between 1·3 and 2·1 times on different occasions. It was therefore routinely added in the work described here. The leakage of plastocyanin from hypotonically disrupted chloroplasts was reported by Katoh and Takamiya (1965) and its role in electron transport to chlorophyll A when ascorbate is used as electron donor has been demonstrated by several workers; cf. Katoh and Takamiya (1963), Davenport (1965) and Vernon and Shaw (1965).

Chlorophyll. When light intensities of 45,000 lux and 5×10^{-6} or 10^{-5} M ferredoxin were present under argon or hydrogen with marrow nitrite reductase in excess, its activity increased with chlorophyll, as marrow grana, beyond 700 μg/ml. The concentration calculated for half maximal rates was 550 μg/ml under these conditions, Fig. 5. When 5×10^{-5} M CMU and the ascorbate-indophenol couple were added with 5×10^{-6} M ferredoxin the half maximal rate was found to require 200 μg/ml chlorophyll. The ratio of $V_{max.}$ for conditions of oxygen evolution to CMU (oxygen evolution inhibited) was about 0·42 whereas in similar experiments there were no differences in the rates of NADP reduction (Losada *et al*., 1961; Hewitt and Betts, 1963).

Light intensity. Increasing light intensity had different effects on nitrite and hydroxylamine reduction by spinach preparations over the range 400–

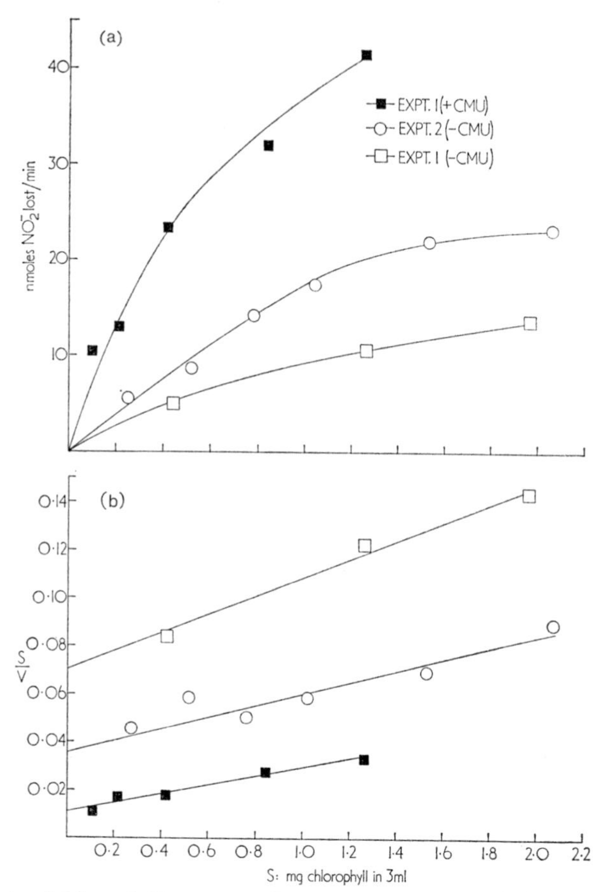

FIG. 5. Effects of chlorophyll concentration as marrow grana and of presence or absence of CMU on activity of marrow nitrite reductase. (a) chlorophyll/activity curves for separate experiments in absence of CMU with preparations of different activity and for presence of CMU in one of these experiments. (b) reciprocal plots of chlorophyll/velocity against chlorophyll. Exp. 1: 240 μg, Exp. 2: 400 μg ferredoxin preparation.

70,000 lux, Fig. 6. Light intensity was saturating for hydroxylamine reductase at the upper limit but not for nitrite reductase. The light intensities producing half maximal rates for nitrite and hydroxylamine reductase activities in the presence of 900 μg chlorophyll and 10^{-4} M ferredoxin were 15,600 and 2900 lux i.e. in a ratio of 5·3 to 1. At light intensities below 5000 lux hydroxylamine was reduced more rapidly than nitrite and at higher intensities the relationship was reversed. The saturating light intensity for NADP reduction by spinach chloroplasts was found to be about 23,000 lux under quite similar experimental conditions (Davenport, 1965).

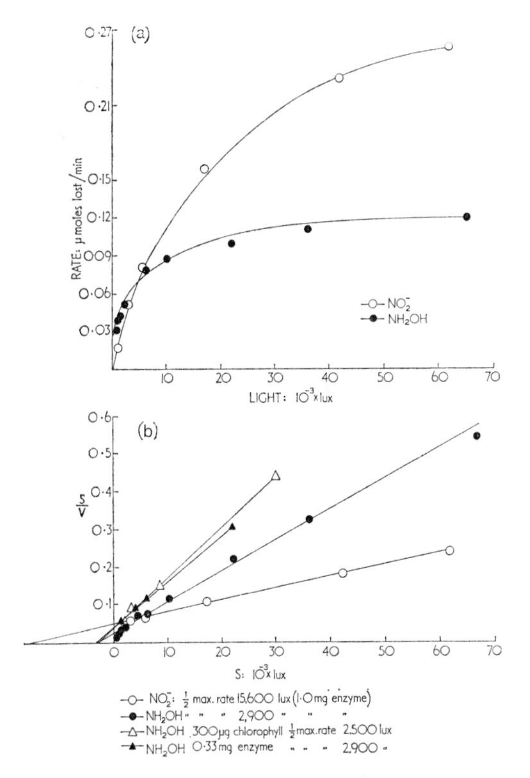

FIG. 6. Effect of light intensity on activity of nitrite and hydroxylamine reductase from spinach. (a) light/activity relationships for each system, showing crossing over of curves at an intermediate light intensity and consequent change in relative specific activities. (b) Reciprocal plots of light/velocity against light for curves in (a) and for two other experiments with hydroxylamine reductase in presence of one-third enzyme (▲) or one-third chlorophyll concentrations (△) used in the upper figure (●).

In experiments where either chlorophyll or enzyme were decreased to one-third of the respective concentrations the rate of hydroxylamine reduction was decreased by about 40% at saturating light intensity (25,000 lux) but the intensity for half maximal activity was practically unchanged, 2500 and 2900 lux for these respective sets. In other experiments described later in relation to effect of nitrite concentration it appeared that substrate inhibition was related to light intensity at low chlorophyll and ferredoxin levels.

Ferredoxin. Purified spinach ferredoxin free of NaCl was used with a spinach enzyme, 45,000 lux and 250 μg chlorophyll/ml as grana. The effects of concentration are shown in Fig. 7 as reciprocal plots. The K_m (in terms of

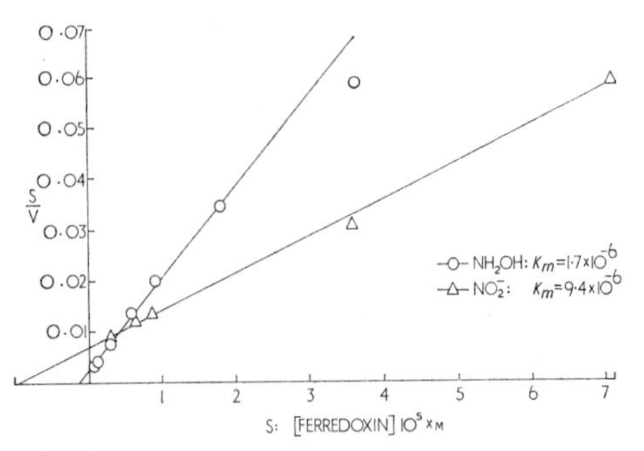

FIG. 7. Effect of ferredoxin concentrations on activity of spinach nitrite and hydroxylamine reductase activities. Plotted from saturation–activity curves of Betts and Hewitt (1966).

total ferredoxin) for nitrite reductase was $9{\cdot}4 \times 10^{-6}$ M and that for hydroxylamine reductase was $1{\cdot}9 \times 10^{-6}$ M. These values are in a ratio of 5 to 1 and similar to the corresponding ratio obtained for light intensity which controls ferredoxin reduction. The K_m for nitrite reductase already reported by Betts and Hewitt (1966) was the same as that obtained by Ramirez *et al.* (1966) using dithionite as the reducing agent. As with the effects of light, the relative activities of nitrite and hydroxylamine reductase were reversed above or below $2{\cdot}8 \times 10^{-6}$ M ferredoxin (Betts and Hewitt, 1966).

Tables I and II show that the photochemical systems at 20°C with 8×10^{-5} M ferredoxin produced much higher nitrite reductase activity than the glucose-6-phosphate system at 25°C with 2×10^{-5} M ferredoxin. The percentage reduction of ferredoxin in the photochemical system is not yet established but if 75%, the ratio of reduced ferredoxin concentrations in the two systems at pH 7·5, and 90% reduction of NADP would be about 20 : 1. In several direct comparisons between the two systems using the same ferredoxin concentrations in both, the rate in the photochemical system was about ten times greater (with plastocyanin) and about six times greater (without plastocyanin) than that in the glucose-6-phosphate system. The rate in the latter was not limited by the potential rate of electron transport to ferredoxin which was 3·5 to 5 times that necessary, and activity was nearly directly proportional to protein beyond the concentration used in such comparisons. The increased activity in the photochemical system was therefore possibly related to the ratio of reduced to oxidized ferredoxin. This may be concerned in an analogous way to that found for BV·/BV by Cresswell *et al.* (1965). Photochemical reduction to 50% or 75% would produce E_0 values of $-0{\cdot}42$ to $-0{\cdot}45$ V whereas in the glucose-6-phosphate system reduction would be about 12% for E_0 value of $-0{\cdot}365$ V (at pH 7·6, 90% reduction of NADP; Mansfield-Clarke, 1960).

Relation to oxygen evolution and atmospheric conditions. Nitrite reduction to ammonia was quantitative under all conditions regardless of oxygen evolution or its suppression by CMU or by heating the grana (5 min at 55 °C), or aerobic or anaerobic atmospheres (Table IV). The maximal rate was obtained under anaerobic conditions with the ascorbate-indophenol couple and heated grana but the advantage was not marked for spinach grana in contrast to the effects of CMU with marrow grana noted above. In the absence of ferredoxin there was no loss of nitrite and as shown in the Methods there was no reduction of nitrite (or of hydroxylamine) by reduced ferredoxin in the absence of enzyme.

Hydroxylamine reduction to ammonia under anaerobic conditions without oxygen evolution was usually quantitative in the presence of ferredoxin but results under other conditions were less reproducible, low recoveries of ammonia being observed on several occasions. When ferredoxin was omitted under aerobic conditions there were large losses of hydroxylamine comparable to usual reductase activity but ammonia recovery was very low. Some oxidative reaction involving hydroxylamine therefore occurred in the absence of ferredoxin but was evidently decreased in activity when ferredoxin was present since ammonia recovery was often 70–90 % under these conditions. There was practically no loss of hydroxylamine in the absence of ferredoxin under anaerobic conditions when oxygen evolution was also suppressed by heating the grana 5 min at 55°C.

SUBSTRATE CONCENTRATION AND INTERACTION

Nitrite concentration. There is considerable discrepancy in reported values for K_m for nitrite and effects of its concentration on activity. Huzisige and Satoh (1961), Joy and Hageman (1966) and Ramirez *et al.* (1966) have reported values of 1 to 3×10^{-4} M for spinach and maize enzymes using either light and grana, dithionite or the glucose-6-phosphate methods to reduce ferredoxin or viologen electron donors. Cresswell *et al.* (1962, 1965) and Sanderson and Cocking (1964) reported values between 1 and $4{\cdot}7 \times 10^{-6}$ M for marrow using BV· or for tomato enzymes using (most probably) endogenous ferredoxin as electron donor. In the present work a range of values has been obtained between $3{\cdot}5 \times 10^{-6}$ and 9×10^{-5} M for the spinach enzyme and ferredoxin reduced either by glucose-6-phosphate or the photochemical systems. The mean value was $3{\cdot}50 \times 10^{-5} \pm 2{\cdot}8 \times 10^{-5}$ M. In several experiments there was inhibition by 2 to 5×10^{-4} M nitrite in agreement with the inhibition above $2{\cdot}7 \times 10^{-4}$ M found for tomato by Sanderson and Cocking (1964). In our experiments with spinach (Fig. 8) there was apparently some degree of substrate co-operation or a biphasic response to increasing nitrite concentration.

TABLE IV

Effect of experimental conditions and presence or absence of ferredoxin on disappearance of nitrite or hydroxylamine, and yields of ammonia produced in different photochemical assay systems with spinach enzymes.

Grana equivalent to 700 μg chlorophyll were prepared as described in the Methods but were not heated unless stated. Anaerobic conditions under argon; oxygen evolution inhibited either by heating grana or adding CMU (5×10^{-5} M) in the presence of ascorbate and dichlorophenol-indophenol. Ferredoxin provided at $2 \cdot 5 \times 10^{-5}$ M. Illuminated 10 min by 40,000 lux. Nitrite or hydroxylamine provided at 2000 nmoles with 0·78 mg protein in 3 ml.

Conditions of assay	Nitrite loss nmoles	Ammonia gain nmoles	Ammonia gain %	Hydroxylamine loss nmoles	Ammonia gain nmoles	Ammonia gain %
Aerobic, complete	1770	1995	113	865	890	102
Aerobic, omitting ferredoxin	50	0	0	445	58	13
Aerobic, heated grana, complete	1785	1920	108	912	530	58
Aerobic, heated grana, omitting ferredoxin	0	0	0	878	114	13
Aerobic + CMU, complete	(I) 1305	1255	96	(I) 898	827	93
				(II) 819	237	29
				(III) 951	733	78
Aerobic + CMU, omitting ferredoxin	0	0	0	807	45	5
Anaerobic, heated grana, complete	(a) 1815	1805	99	(a) 669	603	90
	(b) —	—	—	(b) 225	230	102
Anaerobic, heated grana, omitting ferredoxin	(b) —	—	—	(b) 60	0	0
Anaerobic + CMU, complete	(I) 1645	1640	100	(I) 629	710	113
				(II) 944	850	90

Reciprocal plots showed clearly the existence of a point of inflexion in the curves obtained with the glucose-6-phosphate and ferredoxin system. The discrepancies in reported K_m values cannot be obviously related to the method of electron donation or protein fractionation, neither was any lag period observed in relation to such a possibility raised by Joy and Hageman.

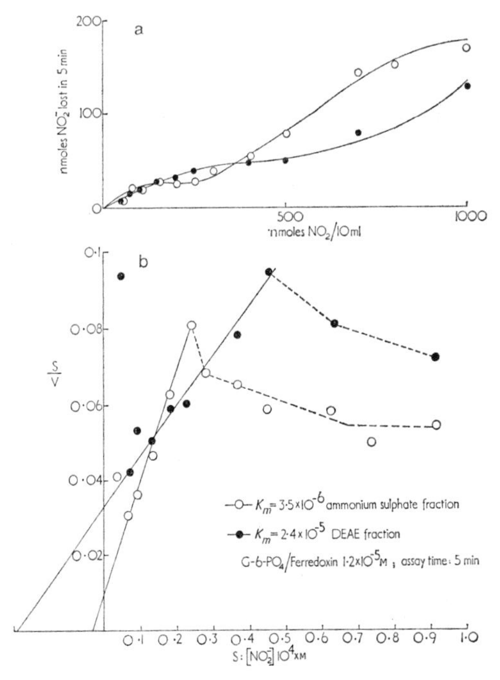

FIG. 8. Relationship between substrate concentration and nitrite reductase activity of two separate spinach preparations assayed with ferredoxin in the glucose-6-phosphate and NADP system. (a) Saturation-activity curves. (b) Corresponding reciprocal plots of S/V against S.

The evidence of substrate co-operation here suggests that the configuration of the protein is a factor. The precise order of adding reactants, the periods during which they are in contact and the atmospheric conditions employed before starting the reaction may be involved in accordance with observations of Ferdinand (1966) on the subject. With high light intensities and ferredoxin concentrations there does not seem to be appreciable inhibition by 7×10^{-4} M nitrite. Results obtained in four experiments using the photochemical system are shown in Fig. 9 where the same enzyme preparation was used in three different light intensities with some other lesser variables. There appears to be some relationship between energy input and substrate-activity kinetics.

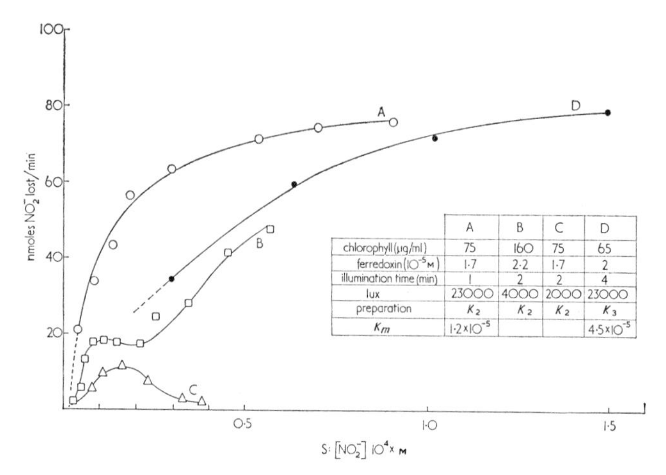

	A	B	C	D
chlorophyll (µg/ml)	75	160	75	65
ferredoxin (10^{-5} M)	1·7	2·2	1·7	2
illumination time (min)	1	2	2	4
lux	23000	4000	2000	23000
preparation	K_2	K_2	K_2	K_3
K_m	$1{\cdot}2 \times 10^{-5}$			$4{\cdot}5 \times 10^{-5}$

FIG. 9. Effects of nitrite concentration on activity of spinach nitrite reductase when assayed with grana and light of different intensities. Experiments A and C were performed on the same occasion, and B and D on separate occasions within a few days. The preparations, K_2 used for experiments A, B and C and K_3 used for D were eluted in adjacent fractions from DEAE cellulose and were drawn from replicate frozen samples.

Hydroxylamine concentration. Substrate cooperation or biphasic effects of hydroxylamine concentration were observed with both marrow and spinach enzymes, Figs. 10 and 11. This phenomenon occurred with BV· (Cresswell *et al.*, 1965) and with ferredoxin and FMN. As ferredoxin is practically specific for the HR1 fraction the effect is not likely to be the result of complex curves due to activities of HR1 and HR2 fractions together. Zero time calibrations carried out for several intermediate concentrations of hydroxylamine are linear and show good precision well within the limits expected for variations due to error, and the experimental results are considered to be valid. Reciprocal plots, where possible, indicate K_m values between 7×10^{-5} and $2{\cdot}4 \times 10^{-4}$ M over the first phase for the spinach enzyme with ferredoxin as electron donor. For marrow with FMN, values up to 4×10^{-4} M have been obtained.

Substrate interaction, Table V. In systems where both nitrite and hydroxylamine are present together, hydroxylamine reduction is markedly inhibited, regardless of the electron donor system, so that HR1 is affected, and HR2 may or may not be. Nitrite reduction is not appreciably inhibited by hydroxylamine under any of these conditions. The K_m for nitrite with BV· was not appreciably changed for the marrow enzyme (Cresswell *et al.*, 1965). When FMN was the electron donor (Fig. 11) the K_m for hydroxylamine was increased from $2{\cdot}7 \times 10^{-4}$ to $9{\cdot}9 \times 10^{-4}$ M by the presence of nitrite and the kinetics were changed from biphasic towards Michaelis and

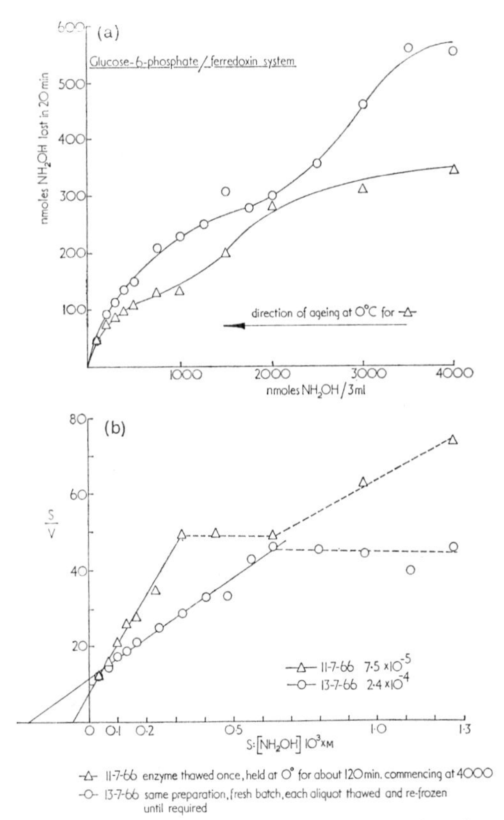

FIG. 10. Relationships between substrate concentration and hydroxylamine reductase activity of the same spinach preparation assayed on separate days with ferredoxin in the glucose-6-phosphate and NADP system but treated slightly differently on the two occasions. (a) Saturation-activity curves. (b) Reciprocal plots of S/V against S.

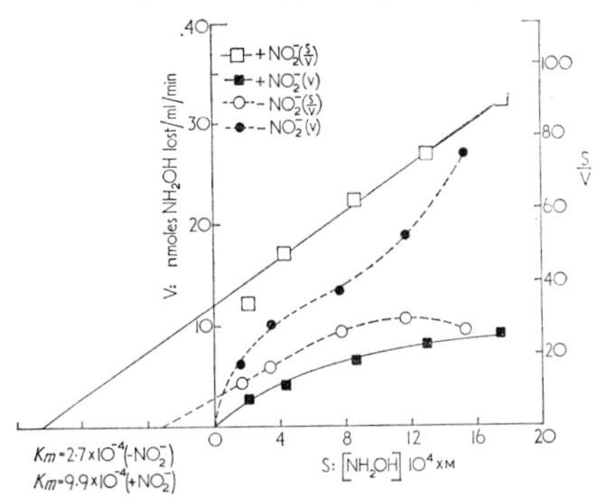

FIG. 11. Relationships between substrate concentration and hydroxylamine reductase activity of a marrow preparation assayed with FMN in the absence or presence of nitrite at 4×10^{-4} M in a reaction volume of 2·15 ml. ■ and ●, saturation activity curves; □ and ○, reciprocal plots of S/V against S.

TABLE V

Inhibition of hydroxylamine reduction by nitrite in marrow enzyme systems.

The electron donor systems were either BV· used in substrate quantities (Cresswell *et al.*, 1965) at two levels of reduction indicated by log BV·/BV, or FMN, or ferredoxin reduced by glucose-6-phosphate and NADP as described in the Methods.

Electron donor system	Substrates provided (nmoles)		Substrate loss (nmoles)		Inhibition % for hydroxylamine
	nitrite	hydroxylamine	nitrite	hydroxylamine	
BV· (log BV·/BV = 0·083)	2000	—	981	—	
	—	2000	—	347	
	2000	2000	985	101	71
BV· (log BV·/BV = $\bar{1}$·301)	2000	—	235	—	
	—	2000	—	295	
	2000	2000	226	20	93
FMN	—	2000	—	637	
	1000	2000	14	36	94
FMN	—	2000	—	724	
	1000	2000	234	343	53
Ferredoxin	2000	—	373	—	
	—	2000	—	472	
	2000	2000	487	210	56

Menten type. This result indicates that nitrite was acting as a modifier for hydroxylamine reductase, probably HR1 which may be closely associated with nitrite reductase activity. In experiments with BV· or FMN represented by the data of Table V there was a more severe inhibition by nitrite when simultaneous nitrite reductase activity was weak. On the other hand, inhibition was decreased only slightly by decreasing nitrite concentrations by a factor of 10 to about 6×10^{-5} M initially. In previous experiments of Cresswell *et al.* (1965), inhibition of hydroxylamine reductase by nitrite was less when nitrite reduction was totally suppressed at BV·/BV ratios below a critical value than when nitrite reduction was also taking place. A rate-limiting intermediate of nitrite reduction may be a common inhibitor for both activities because rate of nitrite reduction as determined by electron donor activity appears to exert more influence than effects of nitrite concentration on the inhibition of hydroxylamine reductase.

DISCUSSION

The results obtained with higher plants show beyond doubt that nitrite is reduced quantitatively to ammonia by highly purified enzymes in a physiological system at rates greatly exceeding those observed with hydroxylamine.

With either ferredoxin reduced under physiological conditions by NADP and glucose-6-phosphate or with BV· the K_m values for hydroxylamine exceed those for nitrite as determined by us. There is no evidence of substrate auto-inhibition for hydroxylamine over the range for which nitrite reduction is the more rapid process. Hydroxylamine is therefore not reduced at comparable or even appreciable rates when present at a fraction of the concentrations at which nitrite reduction occurs rapidly.

Nitrite severely inhibits hydroxylamine reduction even when the latter is present at much higher concentrations, whereas hydroxylamine has a negligible effect on nitrite reduction.

The kinetic data would appear absolutely to exclude free hydroxylamine as an intermediate in nitrite reduction in higher plants. The generalizations of Nicholas (1957, 1958, 1959) argued from less complete kinetic data obtained in experiments with fungi appear therefore not to be applicable to plants.

If a protein-bound intermediate of the same reduction level as hydroxylamine and isomeric with it does occur it cannot be directly in equilibrium with free hydroxylamine when this is added to the system. This conclusion is in full accord with that reached by Lazzarini and Atkinson (1961) from studies with ^{15}N-labelled nitrite added to a system from *Escherichia coli* which was able to reduce nitrite, sulphite and hydroxylamine (Mager, 1960;

Kemp *et al.*, 1963). The sequence proposed for micro-organisms by Fewson and Nicholas (1961) in which nitric oxide and hypothetical nitroxyl are suggested intermediates requires to be tested kinetically. It is the only scheme at present in which $1e^-$ steps are involved and might require the participation of an obligatory $1e^-$ donor.

The problem whether nitrite reductase and the hydroxylamine reductase activity of HR1 are one protein with two related activities or two separate proteins is not yet resolved. The multiple catalytic activity of a probably single enzyme homoserine dehydrogenase—aspartokinase of *E. coli* described by Patte *et al.* (1966) and xanthine oxidase are examples of the former. A tentative hypothetical scheme for reduction of both substrates by a single protein is shown in Fig. 12. This is based on a hemiacetal

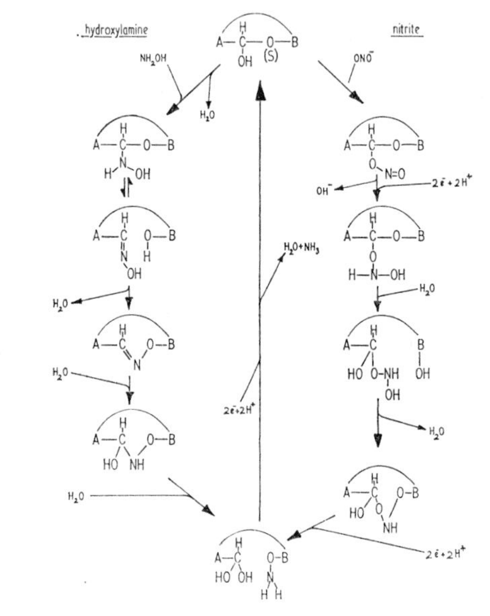

FIG. 12. Tentative scheme for reduction of nitrite or hydroxylamine in a sequence of $2e^-$ steps by the same enzyme system having a hemi-acetal (or hemi-thioacetal (S)) hypothetical prosthetic site and producing no free intermediates.

structure proposed by Cresswell *et al.* (1965) and was stimulated by the original scheme of Kemp *et al.* (1963) who postulated the existence of an active —OH group in the nitrite, sulphite and hydroxylamine reductase system of *E. coli*. Our scheme would be compatible with the absence of free intermediates and not incompatible with the kinetics for substrate and electron-donor concentrations and substrate interactions. Such a protein might also show substrate-cooperation effects if the configuration of the reactive group were dependent on the identity of the substrate and affinity for a given substrate e.g. nitrite could be variable. At present the scheme is based on a sequence of $2e^-$ steps and will need further modifications to accommodate the almost specific requirement for a $1e^-$ donor: either ferredoxin (Tagawa and Arnon 1962) or BV· (Michaelis and Hill 1933a, b) for at least two steps in the reduction of nitrite to the level of the last stage postulated here. It is also apparently necessary to postulate a step before this stage which is in some way related to or limited by the redox value produced by the $1e^-$ electron donor. The different K_m values for both the electron donors ferredoxin, and BV· (Cresswell *et al.*, 1965) depending on which of the two substrates is present could well occur if protein configuration and affinity for the donor depended on which substrate or derived intermediate was present. Evidence for this idea is strengthened by the work of Davies (1961) which showed that the K_m for NADH in the malic dehydrogenase system of plants depends on the specific keto acid acceptor which is used.

It is possible to eliminate practically all hydroxylamine reductase activity from the most highly purified and sometimes from crude preparations but irreversible changes in configuration could be responsible. The transfer of six electrons by a single protein is remarkable. The most highly purified preparations (900–1250 times) yield ammonia quantitatively at high rates. They also persistently show two closely associated bands for spinach and only very faint slower bands after electrophoresis in acrylamide gel. It is possible that electrophoretic separation may induce a "lesion" in the reaction if closely associated proteins, perhaps "twin" oligomers are involved.

The existence of multiple molecular weight species of hydroxylamine reductase, noted also in *N. crassa* by Leinweber *et al.* (1965) and Siegel *et al.* (1965), raises interesting points. At least one of these (HR1) is associated with nitrite reductase in plants and exhibits substrate cooperation or some form of biphasic concentration relationship which might be a manifestation of conformational isomerism described by Rabin (1967). The kinetics of (probably) HR1 with FMN as donor were modified from this pattern to normal Michaelis and Menten kinetics by an inhibitor of the enzyme; either nitrite or an intermediate of the nitrite reduction reaction. The molecular weight species are suggestive of an oligomeric series which show important differences in electron donor specificity; one major fraction HR2 either having an unidentified natural electron donor or else being an artefact. These features are strongly suggestive of the allosteric proteins described by Monod *et al.* (1963) and Monod *et al.* (1965) and may indicate an unrecognized role for hydroxylamine in metabolic control in green plants.

Apparent substrate cooperation in nitrite reductase could be a manifestation of the same phenomenon in the same protein but for the other substrate. Alternatively, two proteins could be involved but the evidence from purification is against this. An extreme example of an analogous biphasic relationship was described for an aldehyde dehydrogenase in seeds by Oppenheim and Castelfranco (1967) for which two K_m values of $2{\cdot}6 \times 10^{-5}$ and 2×10^{-7} M were deduced. Curzon (1967) and others cited by him have found that caeruloplasmin behaves as though either two sites of different K_m occur in the same protein or that steric effects of combination with the substrate at one of several identical sites change the affinity for combination at the remaining sites. The discrepancies in K_m values for nitrite already noted are probably related in some manner to this type of behaviour.

Reduction of nitrite is totally inhibited by NADP in the photochemical system with marrow grana preparations. Shin and Oda (1966) found a similar inhibition in the spinach system. The K_m for ferredoxin in the diaphorase (NADP reductase) is $3{\cdot}3 \times 10^{-7}$ M (Tagawa *et al.*, 1963), compared with $9{\cdot}4 \times 10^{-6}$ M found for nitrite reductase. NADP will therefore be preferentially reduced and the reaction requires less intense light for maximal activity so that there is the basis for suggesting that nitrite reduction may be inhibited until the NADP/NADPH ratio becomes relatively small. Aerobic cyclic photophosphorylation dependent on cyclic reoxidation of ferredoxin also has a lower K_m ($2{\cdot}3 \times 10^{-6}$ M) for ferredoxin than nitrite reduction so that high utilization of ATP may also impede nitrite reduction. That nitrite does not accumulate in tissues indicates that some feedback control may operate to prevent this from happening. The inhibition by nitrite (as HONO) of CO_2 fixation reported by Hiller and Bassham (1965) may be relevant to the postulated control mechanism.

SUMMARY

The status of hyponitrite and hydroxylamine as possible intermediates in nitrite reduction by plants and fungi is reviewed.

Nitrite reduction to ammonia by physiological systems in plants appears to be independent of hydroxylamine which is excluded on kinetic grounds from being a free intermediate.

Highly purified nitrite reductases were obtained from spinach and marrow with specific activities increased between 970 and 1250 times. Maximal specific activities of 48–57 μmoles nitrite reduced to ammonia/min/mg protein were recorded when a photochemical system was used to reduce ferredoxin. Activity-saturation curves for nitrite were often biphasic and K_m values for nitrite were variable.

Existence of multiple hydroxylamine reductase proteins showing different electron donor specificities, substrate cooperation and modification by inhibitors suggested functions found in allosteric proteins. One of these proteins has not been separated from nitrite reductase. A scheme was presented to account for the kinetic and other features of these related systems based on the activity of a possible single protein. A specific role for hydroxylamine in metabolic regulation may be envisaged.

ACKNOWLEDGEMENTS

We are most grateful to Mr B. A. Notton for the electrophoresis examinations, to Mr D. M. James for assistance with the assays and to Messrs. E. F. Watson and R. Fido for growing the plants; Dr E. J. Skerrett supplied the miniature photoelectric cell and relevant microvoltmeter device.

REFERENCES

Anderson, J. H. (1963). *Analyst, Lond.* **88**, 494.
Anderson, J. H. (1964). *Analyst, Lond.* **89**, 357.
Andrews, P. (1964). *Biochem. J.* **91**, 222.
Andrews, P. (1965). *Biochem. J.* **96**, 595.
Arnon, D. I. (1949). *Pl. Physiol.* **24**, 1.
Betts, G. F. and Hewitt, E. J. (1966). *Nature, Lond.* **210**, 1327.
Clendenning, K. and Gorham, P. R. (1950). *Canad. J. Res.* **28**, 114.
Conway, E. J. (1962). "Microdiffusion Analysis and Volumetric Error", 5th Edn. Crosby, Lockwood and Sons, London.
Cresswell, C. F., Hageman, R. H. and Hewitt, E. J. (1962). *Biochem. J.* **83**, 38P.
Cresswell, C. F., Hageman, R. H., Hewitt, E. J. and Hucklesby, D. P. (1965). *Biochem. J.* **94**, 40.
Cresswell, C. F. and Hewitt, E. J. (1960). *Biochem. biophys. Res. Commun.* **3**, 544.
Curzon, G. (1967). *Biochem. J.* **103**, 289.
Davenport, H. E. (1960). *Biochem. J.* **77**, 471.
Davenport, H. E. (1965). *In* "Non Heme Iron Proteins: Role in Energy Conversion" (San Pietro, A. ed.) p. 115. Sympos. Charles F. Kettering Res. Lab., Yellow Springs, Ohio, Antioch Press, 1965.
Davies, D. D. (1961). *Biochem. J.* **80**, 93.
Duane, W. C. and Krogmann, D. W. (1963). *Biochim. biophys. Acta* **71**, 195.
Ferdinand, W. (1966). *Biochem. J.* **98**, 278.
Fewson, C. A. and Nicholas, D. J. D. (1960). *Nature, Lond.* **188**, 794.
Fewson, C. A. and Nicholas, D. J. D. (1961). *Nature, Lond.* **190**, 2.
Frear, D. S. and Burrell, R. C. (1958). *Pl. Physiol.* **33**, 105.
Fry, K. T. and San Pietro, A. (1963). *In* "Photosynthetic Mechanisms in Green Plants", p. 252. *Pubs natn. Res. Coun., Wash.* No. 1145.
Hageman, R. H., Cresswell, C. F. and Hewitt, E. J. (1962). *Nature, Lond.* **193**, 247.
Hewitt, E. J. (1966). "Sand and Water Culture Methods Used in the Study of Plant Nutrition", 2nd Edn. Commonwealth Bureau of Horticulture Tech. Comm. 22. Commonwealth Agric. Bureau Farnham Royal.
Hewitt, E. J. and Betts, G. F. (1963). *Biochem. J.* **89**, 20P.
Hewitt, E. J. and Hucklesby, D. P. (1966). *Biochem. biophys. Res. Commun.* **25**, 689.

Hewitt, E. J. Hucklesby, D. P. and Betts, G. F. (1966). *Biochem. J.* **100**, 54P.
Hewitt, E. J. and Nicholas, D. J. D. (1964). *In* "Modern Methods of Plant Analysis" Vol. VII p. 67 (Linskens, H. F., Sanwal, B. D. and Tracey, M. V. eds.) Springer-Verlag, Heidelberg.
Hiller, R. G. and Bassham, J. A. (1965). *Biochim. biophys. Acta* **109**, 607.
Huzisige, H. and Satoh, K. (1960). *Biol. J. Okayama Univ.* **6**, 71.
Huzisige, H. and Satoh, K. (1961). *Bot. Mag., Tokyo* **74**, 178.
Huzisige, H., Satoh, K., Tanaka, K. and Hayasida, T. (1963). *Pl. Cell Physiol. Tokyo* **4**, 307.
Joy, K. W. and Hageman, R. H. (1966). *Biochem. J.* **100**, 263.
Katoh, S., Shiratori, I. and Takamiya, A. (1962). *J. Biochem. Tokyo* **51**, 32.
Katoh, S. and Takamiya, A. (1963). *In* "Photosynthetic Mechanisms in Green Plants" *Pubs natn. Res. Coun., Wash.* No. 1145, Washington.
Katoh, S. and Takamiya, A. (1965). *J. Biochem. Tokyo* **58**, 396.
Kemp, J. D., Atkinson, D. E., Ehret, A. and Lazzarini, R. A. (1963). *J. biol. Chem.* **238**, 3466.
Kessler, E. (1953). *Arch. Mikrobiol.* **19**, 438.
Kessler, E. (1955). *Nature, Lond.* **176**, 1069.
Lazzarini, R. A. and Atkinson, D. E. (1961). *J. biol. Chem.* **236**, 3330.
Leinweber, F-J., Siegel, L. M. and Monty, K. J. (1965). *J. biol. Chem.* **240**, 2699.
Losada, M., Paneque, A., Ramirez, J. M. and del Campo, F. F. (1963). *Biochem. biophys. Res. Commun.* **10**, 298.
Losada, M., Paneque, A., Ramirez, J. M. and del Campo, F. F. (1965). "Non Heme Iron Proteins: Role in Energy Conversion" (San Pietro, A. ed.) Sympos. Charles F. Kettering Res. Lab., Yellow Springs, Ohio, Antioch Press.
Losada, M., Whatley, F. R. and Arnon, D. I. (1961). *Nature, Lond.* **190**, 606.
Lowry, O. H., Rosebrough, N. J., Farr, A. L. and Randall, R. J. (1951). *J. biol. Chem.* **193**, 265.
Mager, J. (1960). *Biochim. biophys. Acta* **41**, 553.
Mansfield-Clark, W. (1960). "Oxidation-Reduction Potentials of Organic Systems". Williams and Williams, Baltimore.
Medina, A. and Nicholas, D. J. D. (1957a) *Nature, Lond.* **179**, 533.
Medina, A. and Nicholas, D. J. D. (1957b). *Biochim. biophys. Acta* **25**, 138.
Meyer, V. and Schulze, E. (1894). *Ber.* **17**, 1554.
Michaelis, L. and Hill, E. S. (1933a). *J. gen. Physiol.* **16**, 859.
Michaelis, L. and Hill, E. S. (1933b). *J. Am. chem. Soc.* **55**, 1491.
Monod, J., Changeux, J-P. and Jacob, F. (1963). *J. molec. Biol.* **6**, 306.
Monod, J., Wyman, J. and Changeux, J-P. (1965). *J. molec. Biol.* **12**, 88.
Nason, A. (1962) *Bact. Rev.* **26**, 16.
Nason, A., Abraham, R. G. and Averbach, B. C. (1954). *Biochim. biophys. Acta* **15**, 160.
Nicholas, D. J. D. (1957). *Ann. Bot.* **21**, 587.
Nicholas, D. J. D. (1958). *In* Fourth International Congress of Biochemistry (1957) Vol. XIII Colloquia p. 307, Pergamon Press, 1958. London, New York, Paris, Los Angeles.
Nicholas, D. J. D. (1959). *In* "Utilization of Nitrogen and its Compounds by Plants" (Porter, H. K. ed.) p. 1. Sympos. Soc. Exper. Biol. XIII, Cambridge University Press.
Nicholas, D. J. D., Medina, A. and Jones, O. T. G. (1960). *Biochim. biophys. Acta* **37**, 468.

Oppenheim, A. and Castelfranco, P. A. (1967). *Pl. Physiol.* **42**, 125.
Ornstein, L. and Davis, B. J. (1964). *Ann. N.Y. Acad. Sci.* **121**, 305.
Paneque, A. and Arnon, D. I. (1962). *Pl. Physiol.* **37**, suppl. iv.
Paneque, A., del Campo, F. F. and Losada, M. (1963). *Nature, Lond.* **198**, 90.
Paneque, A., Ramirez, J. M., del Campo, F. F. and Losada, M. (1964). *J. biol. Chem.* **239**, 1737.
Patte, J-C., Truffa-Bachi, P. and Cohen, G. N. (1966). *Biochim. biophys. Acta* **128**, 426.
Patterson, D. S. P. (1964). *Biochim. biophys. Acta* **86**, 405.
Rabin, B. R. (1967). *Biochem. J.* **102**, 220.
Ramirez, J. M., del Campo, F. F., Paneque, A. and Losada, M. (1966). *Biochim. biophys. Acta* **118**, 58.
Reisfeld, R. A., Lewis, O. J. and Williams, D. E. (1962). *Nature, Lond.* **195**, 281.
Ritenour, G. L., Joy, K. W., Bunning, J. and Hageman, R. H. (1967). *Pl. Physiol.* **42**, 233.
Roussos, G. G. and Nason, A. (1960). *J. biol. Chem.* **235**, 2997.
Sanderson, G. W. and Cocking, E. C. (1964). *Pl. Physiol.* **39**, 423.
San Pietro, A. (1963). *In* "Methods of Enzymology" Vol. 6 (Colowick, C. P. and Kaplan, N. O., eds.) p. 439, Academic Press, New York.
San Pietro, A. and Lang, H. M. (1958). *J. biol. Chem.* **231**, 211.
Shin, M. and Oda, Y. (1966). *Pl. Cell Physiol., Tokyo* **7**, 643.
Shin, M., Tagawa, K. and Arnon, D. I. (1963). *Biochem. Z.* **338**, 84.
Siegel, L. M., Leinweber, F-J. and Monty, K. J. (1965). *J. biol. Chem.* **240**, 2705.
Silver, W. and McElroy, W. D. (1954). *Archs. Biochem. Biophys.* **51**, 379.
Spencer, D. and Unt, H. (1965). *Aust. J. biol. Sci.* **18**, 197.
Spencer, D. and Wood, J. G. (1955). *Aust. J. biol. Sci.* **12**, 181.
Tagawa, K. and Arnon, D. I. (1962). *Nature, Lond.* **195**, 537.
Tagawa, K., Tsujimoto, H. Y. and Arnon, D. I. (1963). *Nature, Lond.* **199**, 1247.
Vaidyanthan, C. S. and Street, H. E. (1959). *Nature, Lond.* **184**, 531.
Vanecko, S., and Varner, J. E. (1955). *Pl. Physiol.* **30**, 388.
Vernon, L. P. and Shaw, E. (1965). *Pl. Physiol.* **40**, 1269.
Vernon, L. P. and Zaugg, W. S. (1960). *J. biol. Chem.* **235**, 2728.
Virtanen, A. (1961). *A. Rev. Pl. Physiol.* **12**, 1.
Warburg, O. and Christian, W. (1941). *Biochem. Z.* **310**, 384.
Wood, J. G. (1953). *A. Rev. Pl. Physiol.* **4**, 1.
Zucker, M. and Nason, A. (1955). *J. biol. Chem.* **213**, 463.

Section 1e

Nitrate Reducing Enzymes in Barley

B. J. MIFLIN

University of Newcastle-upon-Tyne, England

Most studies of nitrate reduction in higher plants have been carried out with leaf tissue. However, circumstantial evidence suggests that barley and other roots are also able to reduce nitrate to ammonia (Bollard 1956; Hannay *et al.*, 1959 and Joy, 1967), and there are a few reports in the literature of enzyme activity in cell-free root preparations (Sanderson and Cocking, 1964; Wallace and Pate, 1965). It has been shown in our laboratory that the roots of 7-day-old barley plants grown on nitrate contain levels of nitrate and nitrite reductase approximately equal to those in the leaves (Miflin, 1967). The distribution of these enzymes in the root tissue has been studied.

Barley was germinated, suspended on cheesecloth and grown for 7 days. For 18–24 h prior to harvesting the roots were placed in an aerated solution containing 0·02 M KNO_3, 0·0002 M $CaSO_4$ and 0·1 ppm molybdenum. The roots were harvested and ground in 0·2 M sucrose, 0·3 M mannitol, 0·05 M phosphate pH 7·8, 0·005 M EDTA and 0·001 M glutathione. The resulting homogenate was then passed through fine nylon mesh and centrifuged according to the procedure given in Fig. 1. The nitrate and nitrite reductase activities of the various fractions were assayed and the results are shown in Fig. 2. Approximately 20% of the NADH-nitrate reductase and 40% of the nitrite reductase activity was associated with fraction *E*. This fraction is believed to contain the mitochondria. Succinate was also able to donate electrons to nitrate reductase. The distribution of succinate-nitrate reductase presumably follows the distribution pattern of succinic dehydrogenase as well as nitrate reductase. The results show that the total recovery of succinate-nitrate reductase was greater than 100%. This is believed to be due to the greater competition of oxygen as an alternative electron acceptor in the more dilute suspension of mitochondria in the original homogenate as compared with fraction *E*. Succinate does not appear to donate its electrons *via* NAD but more probably *via* a flavoprotein system. Evidence obtained by use of 2-*n*-heptyl-4-hydroxyquinoline-*N*-oxide suggests that a cytochrome b is not involved. At present no donor system other than dithionite and viologen dyes has shown any ability to reduce nitrite,

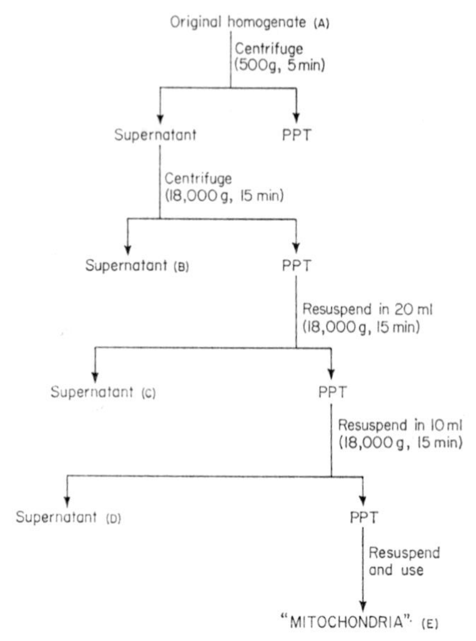

FIG. 1. Fractionation procedure for barley roots.

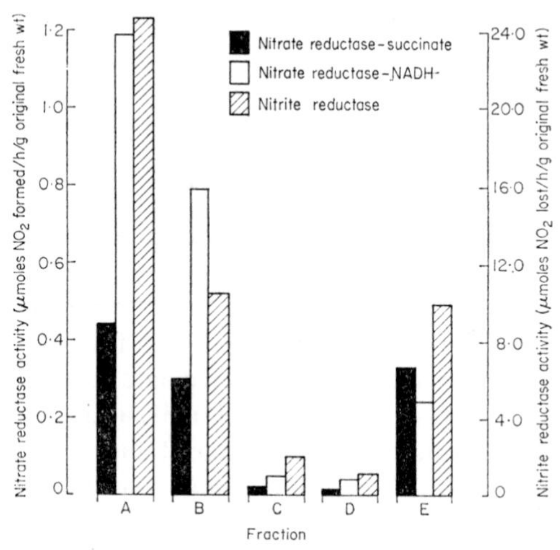

FIG. 2. Distribution of nitrate and nitrite reductase in the subcellular fractions of barley roots.*

* Nitrate reductase assay; 10 μmoles KNO_3, 10 μmoles phosphate buffer pH 7·8, 0·4 ml suitably diluted extract and either 0·27 μmoles NADH or 10 μmoles sodium succinate, total volume 1 ml.

Nitrite reductase assay: 0·5 μmoles $NaNO_2$, 0·25 mg methyl viologen, 1·6 mg $Na_2S_2O_4$, 1·6 mg $NaHCO_3$, 40 μmoles phosphate buffer pH 7·0, and 0·4 ml suitably diluted extract, total volume 1 ml. Fractions explained in Fig. 1.

although so far ferredoxin has not been tried. One of the questions we hope to answer is "what is the physiological reducing system for nitrite in the roots?"

It is possible that the association of these enzymes with the mitochondria is an artefact caused by adsorption of the proteins on to the mitochondria during isolation. To check this, a crude nitrate and nitrite reductase preparation was made by grinding "induced" roots in 0·05 M phosphate (pH 7·8), 0·005 M EDTA and 0·001 M glutathione, and centrifuging off the debris. To this solution was added sucrose (to give 0·2 M), mannitol (to give 0·3 M) and polyvinylpyrrolidone (to give 1 %) and then a mitochondrial suspension prepared from roots in which nitrate and nitrite reductase had not been induced. The mitochondria were then re-isolated using the above procedure (Fig. 1). The original and re-isolated activity of the supernatant and mitochondria was determined (Table I). There was no evidence of

TABLE I

Distribution of nitrite and nitrate reductase in re-isolated mitochondria.

	Total activity/fraction in μmoles NO_2 lost (or formed)/h	
Fraction	NADH-nitrate reductase	Nitrite reductase
Supernatant		
Original activity	10·4	284
Re-isolated activity	8·3	262
Mitochondria		
Original activity	0·0	13·2
Re-isolated activity	0·1	12·8

Assays as for Fig. 2.

adsorption of the enzymes on to the mitochondria because less than 1 % of original supernatant nitrate reductase and none of the nitrite reductase activity was found associated with the mitochondria. The mitochondria were also assayed for succinate-nitrate reductase, but no activity was found despite the fact that the mitochondria were capable of actively oxidizing succinate. The conclusion would therefore appear to be that there is a partial but genuine association of the nitrate reducing system with the mitochondrial fraction of barley roots.

REFERENCES

Bollard, E. G. (1956). *Nature, Lond.* **178**, 1189.
Hannay, J. W., Fletcher, B. L. and Street, H. E. (1959). *New Phytol.* **58**, 142.
Joy, K. W. (1967). *J. exp. Botany* **18**, 140.
Miflin, B. J. (1967). *Nature, Lond.* **214**, 1133.
Sanderson, G. W. and Cocking, E. C. (1964). *Pl. Physiol.* **39**, 416, 423.
Wallace, W. and Pate, J. S. (1965). *Ann. Bot.* **29**, 655.

Section 1f

Mechanisms Involved in the Regulation of Nitrogen Assimilation in Micro-organisms and Plants

A. P. SIMS*, B. F. FOLKES* AND (IN PART) A. H. BUSSEY†

Department of Botany, University of Bristol, England

INTRODUCTION

Over the past two decades much evidence has accumulated to suggest that essentially the same patterns of chemical transformations are used by both plants and micro-organisms to convert the various sources of inorganic nitrogen into the wide range of organic nitrogenous compounds found in living cells. However, while the pathway of nitrogen assimilation appears to have been widely established, there is little information as to the means by which this process is regulated. Studies in recent years on other biochemical pathways have indicated that metabolic control is centred upon enzymes which have both catalytic and regulatory function, these two functions being associated with separate sites within the enzyme structure (Monod *et al.*, 1963). During the course of evolution selective pressure has apparently ensured that the most efficient mechanism of chemical transformation is retained, because the catalytic properties of enzymes from widely different organisms are found to be essentially similar. It seems, however, that the same restrictions have not applied with regard to the regulatory properties of enzymes since evolutionary pressure has here led in different organisms to modifications in both the mechanism and response characteristics of the control sites to enable the enzyme to function in a regulated manner under the particular range of environmental conditions in which the organism is found. In discussing regulatory mechanisms therefore, it is necessary to specify the organisms concerned.

Much of our own work in this field has been with the food yeast *Candida utilis*, a heterotrophic micro-organism generally associated with environments in which a wide range of nitrogen compounds abound. More recently we have devoted some interest to higher plants, particularly the duckweed *Lemna minor*, in the anticipation that their cellular differentiation and adaptation to environments with more restricted availability of nitrogen are likely to be accompanied by some differences in regulatory mechanisms.

* Present address: School of Biological Sciences, University of East Anglia, Norwich.

† Present address: Department of Biological Sciences, Purdue University. Lafayette, Indiana, U.S.A.

REGULATORY MECHANISMS IN FOOD YEAST

A salient feature of the nitrogen metabolism of *Candida utilis* is its ability to grow on a wide range of individual compounds (e.g. nitrate, nitrite, ammonia, urea, many single amino acids and amides) as sole source of nitrogen. The fact, moreover, that its division rate is almost identical on any of these, serves to emphasize the efficiency and flexibility of the underlying regulatory mechanisms in controlling both the rate and nature of the subsequent chemical transformations. Transfer of yeast from one nitrogen source to another in most cases results, after the briefest of pauses, in a rapid resumption of growth, thus indicating that only minor adjustments in the mechanism are necessary to meet the new situation. Further evidence for the flexibility of the control comes from experiments in which both amino acids and inorganic nitrogen are provided together; here both sources are used but the growth rate remains unaltered and hence each compound must alter the rate of utilization of the other, presumably by an effect upon the enzymic mechanisms. Clearly the elucidation of these mechanisms must rest on other more direct experimental approaches, but correspondingly, no description of individual enzymic controls can be deemed adequate unless it is consistent with these broader physiological observations.

The main pathways by which inorganic nitrogen is assimilated in *Candida* are indicated in Fig. 1. In this scheme the central position of ammonia is

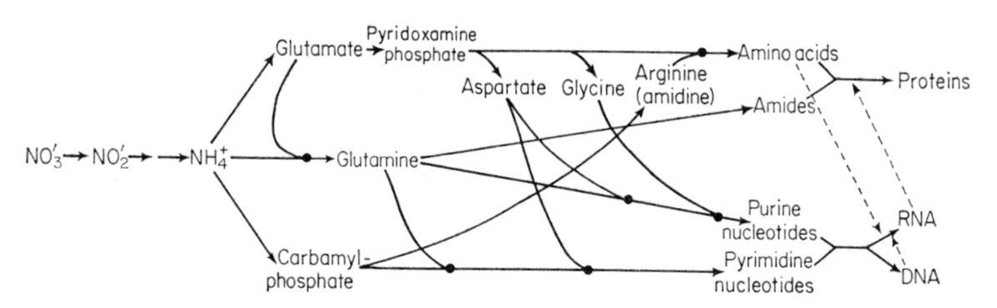

FIG. 1. The pathways of nitrogen assimilation in *Candida utilis*. ———→, pathway of nitrogen; – – →, interaction effect.

apparent, all nitrogen from nitrate passing through this compound before entering organic combination (Sims, 1958). It is only at ammonia that branching of the metabolic network occurs, leading, largely through separate pathways, to the synthesis of proteins and nucleic acids. The importance of glutamate synthesis as the sole source of amino nitrogen in food yeast was first shown by earlier studies in this laboratory using the nitrogen isotope ^{15}N (Sims and Folkes, 1964), and the pathways of nucleotide synthesis from carbamyl phosphate and glutamine have been established in other yeasts (see Buchanan and Hartman, 1959; Reichard, 1959). In an attempt to assess the contribution made by these various pathways we have recently made measurements of the flow along two of them using improved isotopic techniques, and have found that between 75% and 80% of the total nitrogen of the yeast is assimilated as glutamate, and between 10% and 12% as glutamine amide. Since analyses of the content of nucleotides and arginine in the yeast suggest that 6–7% of the total nitrogen must be derived from carbamyl phosphate, it seems that these three pathways are likely to be the only routes of nitrogen assimilation in this organism.

The regulation of this system is of particular interest, because while it represents a series of branched biosynthetic pathways of the type discussed by Stadtman (1966) there are important differences in that none of the branches are truly independent. At later stages all three pathways are linked, both by interchange of substrates and by other effects, such as the amino acid dependence of nucleic acid synthesis (Kurland and Maaløe, 1962; Neidhardt, 1966). Most of the branched pathway systems reviewed by Stadtman (1966) are concerned with the synthesis of a limited number of end products and regulation rests upon the effects of these end products upon the activity and synthesis of the enzyme initiating the pathway. When the initial steps of nitrogen assimilation are considered in this light, however, it is apparent from the complexity of the network that many nitrogenous compounds can be regarded as derived, at least in part, from any particular assimilatory reaction. In such a situation it seems that two possible types of control could arise; regulation could either be delegated to a few key intermediates or else could be by summation of the relatively small effects of each of the many end products. The latter mechanism probably offers the greater flexibility for an organism which may be exposed to a wide range of exogenous substrates; certainly, as the following pages will show, it appears to be the one adopted in food yeast.

FEEDBACK INHIBITION CONTROL OF THE PATHS OF AMMONIA UTILIZATION

A convenient way of studying the feedback mechanisms controlling the assimilation of ammonia in *C. utilis* is by following the effects of a brief period of nitrogen shortage on the subsequent rates of ammonia uptake when the supply is restored (Fig. 2). For our own experiments we have used a turbidostat (Folkes, Sims, Stanley and Wurzberger, see Stanley, 1964), in which a constant cell density can be maintained while allowing the composition of the culture medium to be varied. The advantage of this is that, under a given set of cultural conditions, the yeast goes into a characteristic steady state and this can provide a reproducible starting point for replicate experiments; moreover, the maintenance of a constant cell concentration enables the ready expression of results on a "per cell" basis.

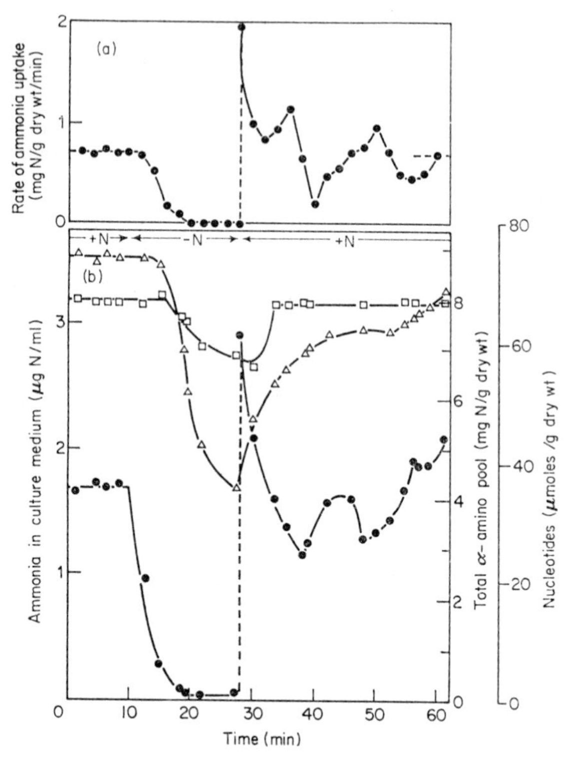

FIG. 2. The effect of a temporary shortage of nitrogen upon the subsequent rate of ammonia assimilation in *Candida utilis*. For a general description of the experiment see below. Cells were cultured at 25°C at a constant density of 0·43 mg dry wt/ml on an inflow medium containing glucose and mineral salts, including 50·8 μg ammonium–N/ml. At 10 min the turbidostat inflow was switched to a nitrogen deficient medium; at 28 min 5·9 mg ammonium–N was added as NH_4Cl to the 2 litres of culture and simultaneously the inflow was switched back to the original medium.
(a) ●, the rate of uptake of ammonium-N by the yeast cells, as calculated from the changes in ammonium level in the culture and the rate of inflow of fresh medium. (b) ●, residual ammonium-N in the culture medium; △, the level of α-amino-N associated with the free amino acids of the yeast cells; □, the level of free nucleotides in the yeast.

Cells were cultured on an inflow medium containing just sufficient ammonium phosphate to maintain a small but measurable excess of ammonia in the culture. When the cells had reached a steady state under these conditions, the inflow medium was changed to one in which ammonia was absent. The level of free ammonia in the culture medium then fell rapidly. After a few minutes of this nitrogen depletion sufficient ammonia was added to the culture vessel to more than restore the original ammonia level and at the same time the inflow was switched back to the original ammonia medium. In spite of this restoration the level of free ammonia in the culture again fell rapidly, indicating that the rate of ammonia uptake had been increased to well above the steady-state level as a result of cellular changes occurring during the brief period of nitrogen shortage. Such rates in excess of the steady-state reveal a potential for catalytic activity which must normally be kept under restraint; a restraint which is itself removed during the period of nitrogen depletion. Calculations showed that the rate of ammonia uptake later fell and approached the original steady-state value by a series of progressively damped oscillations. This suggests that control was by then once more being applied. Some indication of the mechanism of this regulation came from an analysis of the major soluble nitrogenous components within the cells. These end products of ammonia assimilation fell markedly during ammonia deprivation so that the highest rates of ammonia uptake coincided with the lowest levels of the amino acid and nucleotide pools; moreover, the subsequent return to the original rate of uptake occurred only when these pools had been restored once more to their steady-state levels.

From data of this type it is not possible to do more than show the likelihood that end products of nitrogen metabolism affect, by feedback control, the rate of ammonia assimilation. Fuller understanding can come only from a study of the effects of end products upon the individual pathways of ammonia utilization; by using the rate of incorporation of isotopic nitrogen to measure the changing rates of assimilation along the various pathways, it is possible to investigate their control in living cells.

Of the three major routes of ammonia incorporation, that leading to the synthesis of glutamic acid has been most fully investigated. As earlier shown by us, this is the step by which all α-amino nitrogen is synthesized and hence all amino acids may be considered to be ultimate end-products of this pathway. Isotopic measurement of the rate of glutamate synthesis during the ammonia restoration phase of a depletion experiment such as described above shows that the exceptionally high initial rate of amino acid synthesis soon falls off in a characteristic way which is closely mirrored by the rising level of the total pool of amino acids within the cells (Fig. 3). The extremely close correlation between rate and total α-amino level holds over a three-fold change in rate and is apparently independent of the level of individual amino acids and of other soluble nitrogenous compounds. Evidence that the regulation of glutamate synthesis is through the inhibitory action of the amino acid end-products has come also from studies on the isolated glutamic acid (NADPH) dehydrogenase. All amino acids, either alone or in combination, are able to inhibit; those acids, including glutamine, which are major components of the cellular pool being particularly effective (Table I). As with the *in vivo* response shown in Fig. 3, the enzyme

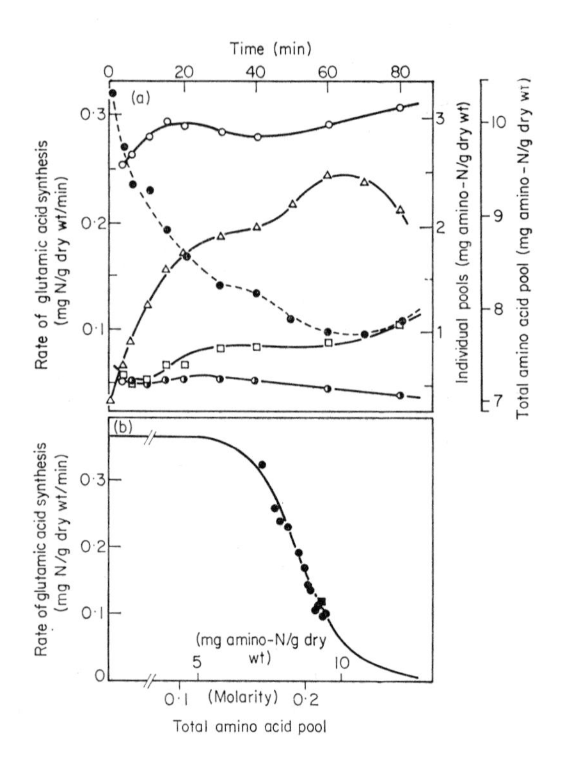

FIG. 3. The regulation of glutamic acid synthesis *in vivo* as revealed by isotopic incorporation following a temporary shortage of nitrogen. The experimental procedure was similar to that for Fig. 2 except that culture was at 15°C and the restored nitrogen supply contained ^{15}N ammonium phosphate. Times are in minutes after this restoration. (a) ●, rate of glutamic acid synthesis as calculated from the rate of ^{15}N incorporation; △, the level of the total pool of free amino acids in the yeast; ○, free glutamic acid; □, free glutamine; ◑, free aspartic acid. (b) The correlation between rate of glutamate synthesis and the concentration of the amino acid pool. ●, the values based on the data shown above; ■, point derived from a separate experiment at 15°C in which glutamate synthesis was measured in cells growing in steady state on a constant ammonia supply. The line drawn through the points is that for a feedback inhibition which is proportional to the eighth power of amino acid concentration.

in vitro exhibits cooperative effects of increasing inhibitor concentration (cf. Monod *et al.*, 1963). Its behaviour differs however from the cooperative feedback inhibition as defined by Stadtman (1966) in that more than one species of inhibitor does not have to be present in order for cooperative effects to be shown; we propose therefore to describe this as species-independent cooperative inhibition. Unlike the other amino acids, glutamic acid, when present at the concentration at which it occurs in the metabolic pool, has an activating effect upon the enzyme (Fig. 4). A similar effect,

TABLE I

In vitro inhibition of glutamic (NADPH) dehydrogenase of *Candida utilis* by amino acids.

Assays, on an acetone-powder preparation, were in the direction α-oxoglutarate to glutamate under the following conditions: 21°C; in 0·1 M-phosphate buffer at pH 8·3 with 0·008 M-α-oxoglutarate; 0·015 M-NH_4Cl; 0·0002 M-NADPH; L-amino acids were uniformly tested at a concentration of 0·2 M.

Amino acid	Concentration of amino acid in free amino acid pool of yeast (μ mole/ml cell volume)	% inhibition of dehydrogenase by 0·2 M-amino acid
Total pool	220	43
Glutamic acid	76·5	47
Glutamine	32·1	24
Alanine	30·6	31
Arginine	26·4	50
Aspartic acid	10·9	27
Lysine	8·8	48
Serine	7·8	35
Histidine	4·8	51
Threonine	4·4	31
Glycine	4·4	17
Citrulline	3·5	22
Valine	3·0	11
Ornithine	2·2	27
Leucines	0·5	N.D.*
Tyrosine + phenylalanine	0·5	N.D.

Similar but less extensive results have been obtained with a highly purified dehydrogenase derived from a hammer-press preparation and the percentage inhibition by 0·2 M amino acids was considerably higher.

* N.D., not determined.

which is probably important in the coordination of carbon metabolism and nitrogen metabolism, is activation by low levels of α-oxoglutarate. With this exception the pathway appears to be solely under end-product control since no effect of products of the parallel pathway leading to carbamyl phosphate and nucleotides can be detected either *in vivo* or *in vitro*.

The biological advantages of this mechanism, in which all amino acids function in a feedback control, rest not only on its ability to regulate the rate of amino acid synthesis for changing needs but also on its flexibility

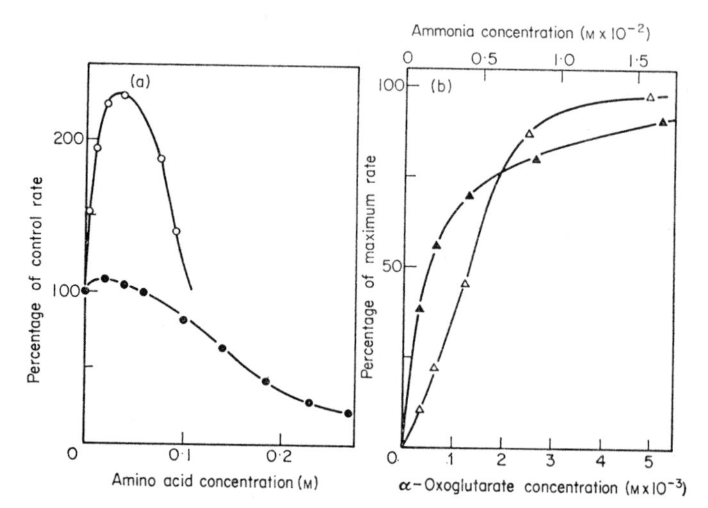

FIG. 4. Allosteric effects on glutamic (NADPH) dehydrogenase from yeast (unpublished data of Rowena Jones).
(a) Effects of varying concentrations of two amino acids, expressed as a percentage of the reaction rate in the absence of the effector. ○, glutamic acid; ●, arginine. (b) Effects of increasing substrate concentration on the reaction rate, expressed as a percentage of the maximum rate. △, α-oxoglutaric acid; ▲, ammonium phosphate. Note that the curve for α-oxoglutarate is not of simple hyperbolic form comparable to that for ammonia; instead the rate of enzyme reaction shows relatively little response to changing substrate at low substrate concentrations.

of response so that any one amino acid, when supplied together with ammonia, can spare the expenditure of carbon and energy in glutamate synthesis. The very sharp response of rate of synthesis to changes in pool level means that these effects can be brought about by relatively small alterations in amino acid concentration, an important consideration if the rates of amino acids utilization in, for example, protein synthesis are not also to be affected. Under normal steady-state conditions the pool of amino acids is relatively large, so that there is adequate provision for increased enzymic activity if required; situations in which this may be necessary include conditions of low ammonia supply, when substrate limitation may affect enzymic efficiency, and growth at higher temperatures, when the low Q_{10} of the enzyme as compared with that for growth means that amino acid utilization would outstrip synthesis were it not for this control. In both of these cases quantitative experiments show that the feedback mechanism alone is capable of adjusting synthesis to requirement.

The isotopic approach has also been used in the study of the glutamine pathway, and here, in spite of the current belief that all branch points in metabolic pathways are likely to be subject to feedback control, we have

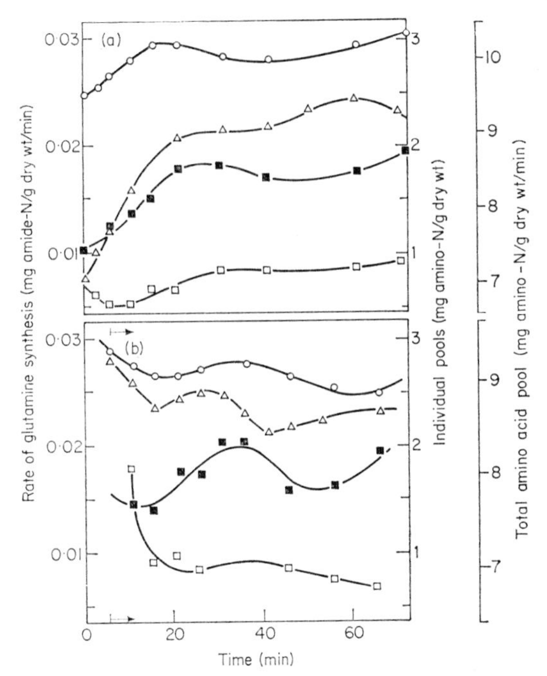

FIG. 5. Changes in the rate of glutamine synthesis and associated fluctuations in the level of pool components during departures from steady state.
(a) Changes immediately following the restoration of nitrogen to a depleted culture. The data derive from the same experiment as is shown in Fig. 3. (b) Changes associated with raising the temperature of the culture. The culture, growing on an ammonium medium, was in steady state at 15°C. Over the five-minute period indicated ⊢→ the temperature was raised to 21·5°C and maintained at this point. ■, rate of glutamine synthesis calculated from ^{15}N incorporation; □, glutamine pool; ○, glutamic acid pool; △, total pool of amino acids.

been unable to find any evidence for such regulation in yeast cells. No negative correlation between the rate of glutamine synthesis and the concentration of glutamine or of any free amino acid or nucleotide could be detected in the data from a number of experiments (Fig. 5). Instead there are consistent indications that the rate of synthesis is positively correlated with the level of free glutamic acid (see Fig. 6), and, under conditions when glutamic acid is in limited supply, with the level of ammonia.

The apparent lack of feedback control on the synthesis of glutamine raises the question as to how the competition for ammonia between glutamic acid synthesis and glutamine synthesis is regulated. It seems to us that direct feedback control is unnecessary in this case since the interactions between the two assimilatory systems provide another mechanism. This depends upon the fact that glutamate level governs the rate of glutamine

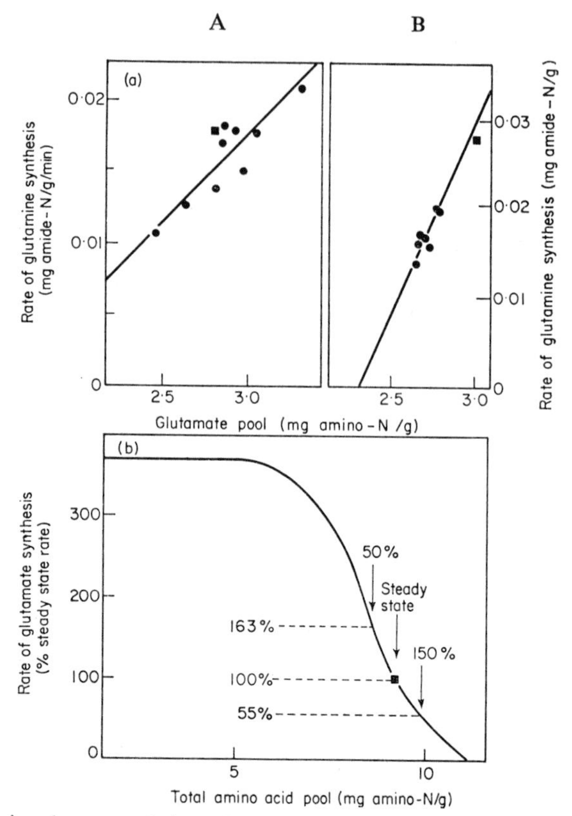

FIG. 6. The involvement of glutamic acid in the regulation of glutamine synthesis. (a) Correlations between the rate of glutamine synthesis and the concentration of free glutamic acid. ●, data derived from the experiments shown in Fig. 5: A, nitrogen depletion at 15°C; B, temperature transition to 21·5°C. ■, data derived from separate steady-state experiments at the corresponding temperature.
(b) The interactions of glutamine synthesis and glutamic acid synthesis. Under or over production of glutamine, dependent upon glutamic acid availability, leads to fluctuations in the glutamine pool which affect the rate of synthesis of glutamic acid and hence its availability. The graph, derived from Fig. 3, shows the effect of variations in the glutamine pool, from 50% to 150% of the steady-state level, on the rate of glutamic acid synthesis.

synthesis while the glutamine level, through negative feedback, regulates the rate of glutamate synthesis. Any over-production of glutamine will lead to increase in the total amino acid pool and hence to reduction in the rate of glutamic acid synthesis; the resulting decrease in glutamic acid concentration in the cell will bring about reduction in the rate of glutamine synthesis. Under-production, by reducing the size of the glutamine pool, will have a reverse effect. This mechanism is practicable because glutamine forms a relatively large proportion (up to 15%) of the amino acid pool; as a consequence reductions in glutamine level of the order of 50%, as have commonly been observed in our nitrogen depletion experiments, result in a large enough decrease in the level of total amino acids to allow a 60% increase in the rate of glutamic acid synthesis (Fig. 6). In other micro-organisms, work with the isolated glutamine synthetase suggests that this may be subject to cumulative feedback control (Stadtman, 1966), but in these organisms glutamine does not represent such a major component of the soluble pool; the situation in yeast appears to make such additional control mechanisms unnecessary.

CONTROLS INVOLVING CHANGES IN THE LEVEL OF ASSIMILATORY ENZYMES

So far, the discussion has been limited to experiments of such short duration that changes in enzyme level were not involved and all regulation could be shown to be by control of the activity of a constant amount of enzyme. In longer experiments however, in which cells are first grown in ammonia and then transferred to a medium containing both amino acids and ammonia, there are two effects: immediately, the glutamic (NADPH) dehydrogenase activity is reduced as a result of the accumulation of amino acid within the metabolic pool, and, more slowly, the actual level of enzyme falls. Many experiments indicate that this enzyme repression is, like the inhibition, brought about by the total pool of amino acids; in the turbidostat over a wide range of cultural conditions a very close negative correlation can be demonstrated between amino acid level and the amount of enzyme/cell.

The significance of this control mechanism lies not only in the economy of enzyme production which results when exogenous amino acids are available, but also in its flexibility to meet the contrasting situation which occurs when the supply of substrate for glutamate synthesis is limiting. Under conditions of reduced availability of ammonia, such as apply when the ammonia level in the culture medium is kept low, or when nitrate or nitrite are the sole sources of nitrogen, the resulting decrease in the total amino acid pool will lead to extensive derepression. The high level of enzyme then enables the yeast to maintain the same rate of growth in spite of the low availability of substrate. This effect is particularly marked in cells grown on nitrate at higher temperatures since the high rate of growth and the low Q_{10} of the nitrate reduction system combine to reduce ammonia availability to a point where the pool level is low enough to give three times the amount of enzyme as is formed on ammonia media (see Fig. 7). Even at lower temperatures when the amino acid pool is considerably greater, there is virtually no feedback control operative upon the dehydrogenase when nitrate is the nitrogen source, so the control of assimilation under these conditions rests on the regulation of the new first member of the

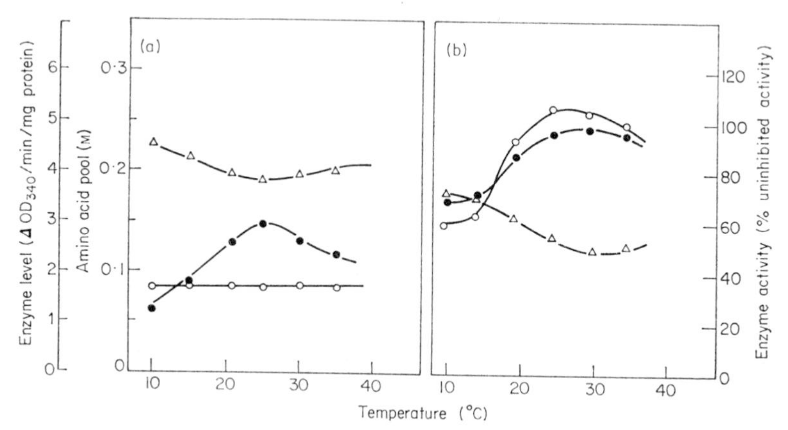

FIG. 7. The effect of the amino acid pool, under varying conditions of temperature and nitrogen source, on the degree of feedback inhibition and repression of glutamic (NADPH) dehydrogenase.
(a) steady state growth with ammonium phosphate. (b) steady state growth with potassium nitrate. △, amino acid pool; ○, enzyme level/mg of cellular protein; ●, activity remaining after inhibition by the amino acid pool, expressed as a percentage of the uninhibited activity (data calculated from the feedback inhibition curve given in Fig. 3).

assimilatory pathway, the nitrate reduction system; in this regulation the amino pool again has a part.

CONTROL OF THE ASSIMILATION OF NITRITE AND NITRATE*

There is now decisive evidence that the sole mechanism for the assimilation of nitrate or nitrite by food yeast involves reduction to ammonia and organic combination along the three pathways already discussed. The nitrate-reducing system must thus satisfy the requirements of nitrogen for both amino acid and nucleotide synthesis; the fact that externally supplied amino acids and nucleotides are utilized in the presence of nitrate suggests the possibility that both of these classes of compounds may have a direct regulatory effect on nitrate assimilation.

Whereas the actual nature of the intermediates involved in the process of nitrate reduction may be a matter of controversy, several facts about the nitrate reduction system of yeast are indisputable; it is possible to isolate an enzyme complex which can, with a specific requirement for NADPH, bring about the almost quantitative conversion of nitrate to ammonia, and this same complex is capable also of reducing nitrite and hydroxylamine at rates, under optimum assay conditions, equal to that for the reduction of nitrate (Bussey and Sims, see Bussey, 1966). No activity with another hypothetical intermediate, hyponitrite, can be detected. The 1 : 1 : 1 ratio of activity for the reduction of the three reactive compounds is maintained

* With A. H. Bussey.

over the very wide range of enzyme level which results from varying cultural conditions, including induction with nitrate (Fig. 8) or nitrite, and repression

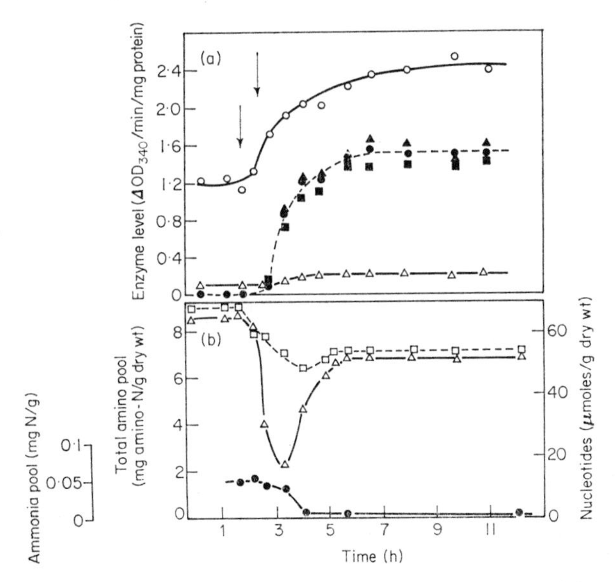

FIG. 8. Co-ordinate induction of the enzymes associated with the nitrosome in yeast (Data of Bussey, 1966). Cells were grown in a turbidostat at 30°C on an ammonium medium and at an appropriate time (indicated by the first arrow) an excess of potassium nitrate was added to the culture vessel and the inflow medium was replaced by one containing nitrate as nitrogen source. It was not until the ammonia in the medium was exhausted (indicated by second arrow) that appreciable changes in the level of enzymes and pool components could be detected.
(a) Changes in enzyme level. ●, nitrate reductase, ▲, nitrite reductase; ■, hydroxylamine reductase; ○, glutamic (NADPH) dehydrogenase (× ½); △, glutamic (NADH) dehydrogenase. (b) Changes in level of soluble intermediates within the yeast cells. □, total nucleotides; △, total amino acids; ●, ammonia.

by amino acids or ammonia. The co-ordinated nature of these enzymic activities has led us to postulate the association of the individual enzymes into an oligomeric complex, of molecular weight about half a million, this nitrosome being capable of bringing about in a controlled manner the reduction of nitrate or nitrite to ammonia without the intervention of free intermediates. The lines of evidence supporting this are summarized below: (*a*) The almost quantitative conversion of small amounts of nitrate to ammonia and the extremely low level of detectable intermediates suggest that all the individual reactions are linked within a single enzyme complex and that equilibration with external pools of intermediates does not occur. This is particularly evident in the case of hydroxylamine since the very large K_m of the hydroxylamine reductase (see Table II) would make high

TABLE II

K_m for substrates of the nitrosome complex.
(Data of Bussey, unpublished).

Substrate	K_m
Nitrate	$1{\cdot}2 \times 10^{-4}$ M
Nitrite	$1{\cdot}5 \times 10^{-5}$ M
Hydroxylamine	$6{\cdot}5 \times 10^{-3}$ M

concentrations of this intermediate obligatory if fast rates of conversion of nitrate to ammonia were to be maintained in a system of unlinked enzymes.

(*b*) Enzyme preparations can be fractionated on agarose columns into a series of components which are capable of reducing nitrate or nitrite to ammonia. The molecular weights of these components range from about 65,000 to over 500,000 (see Table III) and they probably represent an oligomeric series with 65,000 as the smallest functional unit or monomer. It is also possible to detect a single enzyme of lower molecular weight (17,000) which can reduce only nitrate to nitrite: this has modified substrate affinity and regulatory properties and is likely to represent a subunit dissociated from the complex which retains some catalytic activity.

(*c*) All compounds which inhibit the conversion of nitrate to ammonia affect equally the reduction of nitrite and hydroxylamine (Fig. 9). The

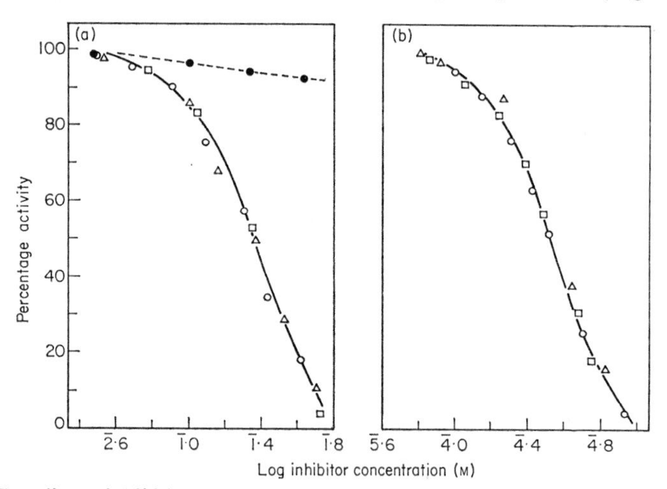

FIG. 9. Coordinate inhibition of the enzymes of the nitrosome (Data of Bussey, 1966). The effect of varying concentrations of (a) ammonia and (b) pyridoxamine phosphate on the enzyme activities, expressed as a percentage of the uninhibited activity.
○, nitrate reductase in nitrosome; △, nitrite reductase; □, hydroxylamine reductase; ●, nitrate reductase released by breakdown of the nitrosome by repeated freezing and thawing.

TABLE III

Molecular sizes of the enzyme aggregates associated with the reduction of nitrate to ammonia.

Enzyme preparations from yeast were obtained by use of a hammer press and were fractionated by gel-filtration on columns of agarose granules. Various agarose concentrations were used in order to allow the resolution of a wide range of molecular sizes; sizes not resolvable at a particular agarose concentration are indicated NR in the table below. Enzymic activity was assayed in many different ways: (a) nitrite formation from nitrate with NADPH; (b) nitrite formation from nitrate with NADH; (c) nitrite-dependent NADPH oxidation; (d) nitrite disappearance in the presence of NADPH; (e) ammonia formation from nitrate; and (f) ammonia formation from nitrite. Molecular sizes associated with these activities were estimated from column calibrations carried out with proteins of known Stokes radius. The molecular weights are calculated on the assumption that the molecular density is that of typical proteins of corresponding Stokes radius.

Assay method	% Agarose	Experiment No.	Stokes radii corresponding to elution peaks				
Nitrate reduction (a and b)	4	1	83·5	60·5	40·5	NR	NR
		2		62·5	46·0	NR	NR
		3			44·0	NR	NR
	8	5	NR	62·5	44·5	33·5	20·5
	10	8	NR	NR		38·0	19·0
		9	NR	NR			18·5*
		10	NR	NR			19·0*
		11	NR	NR			18·5*
Nitrite reduction (c and d)	4	4		61·0	45·0	NR	NR
	8	6	NR	59·5	45·5		
	10	12	NR	NR	47·5	33·0	
Ammonia formation (e and f)	4	1	83·5	60·5	40·5	NR	NR
	8	7	NR	59·5	48·0	33·5	
	10	12	NR	NR		33·0	
Best values			83·5	60·5	45·0	33·5	19·0
Approximate molecular weight			520,000	260,000	130,000	65,000	17,500

* In these preparations dissociation into subunits was brought about by preliminary treatment of the enzyme aggregate with 10^{-2} M glutathione (reduced).

nitrate reductase activity which survives dissociation of the complex is not so affected by inhibitors.

(*d*) The production of all three enzymic activities appears to be obligately linked, even though not required in a particular biological situation. Thus when nitrite alone is supplied, the ability to reduce nitrate still survives although no longer necessary; similarly, under anaerobic conditions when ammonia is also present, nitrate is quantitatively converted to nitrite in a dissimilatory reaction, but nitrite and hydroxylamine reductase activities are still present at levels equal to that of the nitrate reductase.

The control of the level of the enzyme complex per cell involves both positive effectors (inducers) and negative effectors (repressors). Nitrate and nitrite are inducers, and only minute amounts of enzyme are formed in their absence, even under conditions such as nitrogen starvation in which

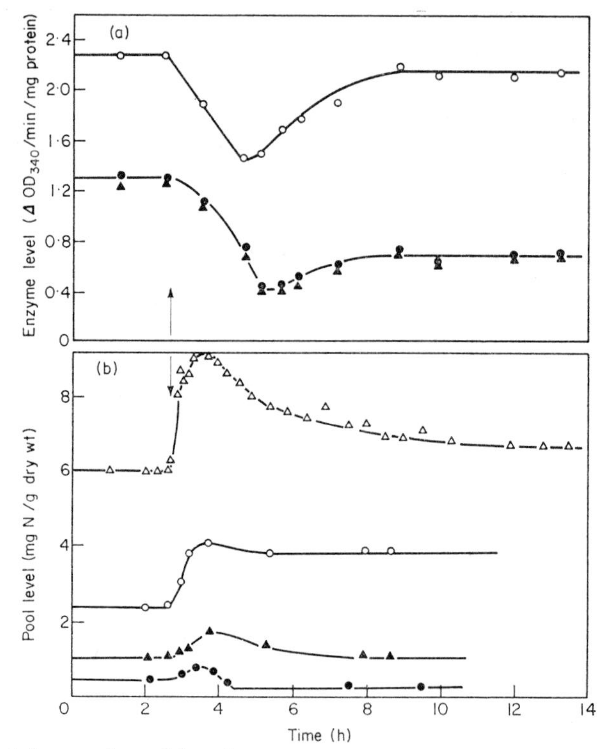

FIG. 10. Partial repression of the nitrosome and other enzymes in the presence of exogenous glutamate (Data of Bussey, 1966). Yeast was grown in a turbidostat at 25·5°C on a nitrite medium and at the point indicated the inflow was switched to a medium containing an excess of both nitrite and L-glutamate.

(a) Changes in enzyme level. ●, nitrate reductase; ▲, nitrite reductase; ○, glutamic (NADPH) dehydrogenase (×½). (b) Changes in pool components within the yeast. △, total amino acids; ○, glutamic acid; ▲, aspartic acid (×2); ●, ammonia (×50).

the concentration of repressors is low. On the other hand, culture with ammonia or amino acids in the presence of nitrate leads to varying degrees of repression of enzyme production. In the presence of ammonia, which can readily supply the total nitrogen requirements of the yeast, this repression is complete, a fact which has led many workers to suggest that this compound is the actual repressor of nitrate reductase. However, our own experiments indicate that this effect is not a direct one but is through the raising of the concentration of amino acids and nucleotides which results from the increased availability of ammonia. In certain circumstances, for example when glutamic acid is added to a culture growing on nitrite (Fig. 10), the enzyme is substantially repressed even though the concentration of free ammonia in the cells falls considerably; the time course of enzyme repression in this experiment, while showing little relation to changes in the level of individual amino acids, correlates remarkably closely with the complex changes in the size of the total amino acid pool. It would seem likely, from the results of sparing experiments, that nucleotides as well as amino acids are involved in the regulation of the nitrate pathway, but we have as yet little direct evidence as to their participation at the level of enzyme repression.

The feedback mechanisms concerned with the control of the enzymic activity of the nitrosome appear to be much more complex than those governing the glutamic dehydrogenase and for this reason study of the system has tended to centre upon an investigation of the inhibition properties of the isolated enzyme. Once these were known it was possible to look for their participation in the overall control in living cells. *In vitro* inhibition can be demonstrated for a wide range of compounds (see Figs. 9 and 11): ammonia and practically all amino acids and nucleotides inhibit to varying extents, and pyridoxamine phosphate and carbamyl phosphate are particularly effective. There is some indication that several different receptor sites are involved in these inhibitions because both competitive and noncompetitive action has been observed. The inhibitory effects of different compounds appear to be cumulative, even though inhibition curves of rate against concentration for individual compounds indicate some degree of allosteric interaction. At the concentrations at which they can occur in cells, amino acids or nucleotides are capable of only partial inhibition of the nitrosome system, but the fact that products, all of the main paths of nitrogen assimilation, are able to act on the enzyme complex makes possible, if the quantitative behaviour of the complex within the cell is the same as in the test tube, a sensitive and versatile mechanism by which the rate of nitrate reduction may be controlled in response to changes in the availability of individual amino acids and nucleotides.

The interactions of these different control mechanisms have been

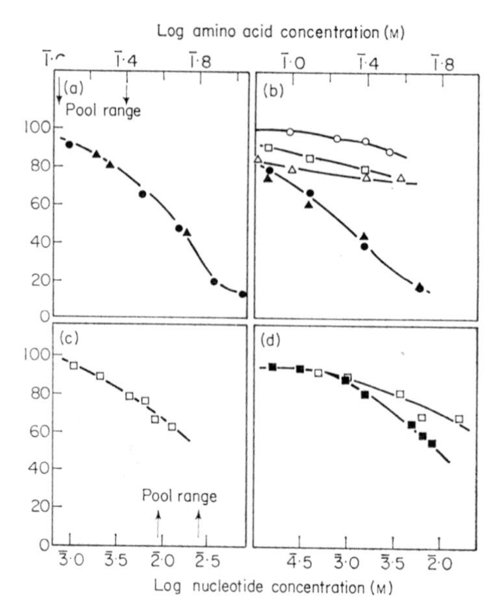

FIG. 11. Inhibition of the nitrosome by amino acids and nucleotides (Data of Bussey, 1966). Inhibited rates are expressed as a percentage of the uninhibited rate; assays are of nitrate reductase activity, except where indicated.
(a) The effect of an amino acid mixture simulating the pool in yeast (see Table I); the normal concentration range *in vivo* is indicated. ●, nitrate reductase; ▲, nitrite reductase. (b) The effect of individual amino acids. ○, aspartic acid; □, glutamic acid; △, glycine; ●, lysine on nitrate reductase; ▲, lysine on nitrite reductase. (c) The effect of a nucleotide mixture simulating the yeast pool (Gilbert, 1959); the normal concentration range *in vivo* is indicated. □, nitrate reductase. (d) The effect of individual nucleotides. □, adenosine triphosphate; ■, nicotinamide adenine dinucleotide.

conveniently studied by experiments in which cells growing on nitrite have been exposed to gradually increasing concentrations of an amino acid or ammonia. When this is done in a turbidostat it is possible to calculate from the changes in concentration with time, the rates of uptake of both nitrite and the alternative nitrogen source and these, together with measurements of the levels of both enzyme and cellular intermediates, provide a comprehensive picture of the interplay of end products and control. The results of a typical experiment, involving competition between nitrite and ammonia, are given in Fig. 12. As soon as ammonia was made available it was immediately utilized and there was a resultant decrease in the rate of nitrite utilization. The falloff in nitrite uptake was reversed for a short period, corresponding to a phase of very rapid cell division, but eventually fell to zero. Associated with this was a marked increase in the cellular pools of amino acids and ammonia. Later, nitrite was again used extensively, and this at a time when repression and destruction of the nitrate reductase had

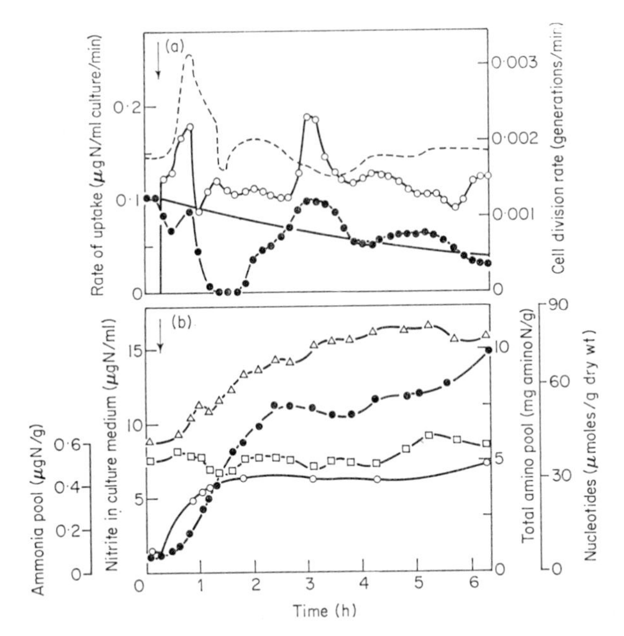

FIG. 12. Factors affecting the activity of nitrosomes *in vivo*, as revealed by response to supplying nitrite-grown yeast with both ammonia and nitrite (based on unpublished data of Bussey). Yeast was grown in a turbidostat at 14·5°C on a medium containing sodium nitrite at a level sufficient to provide a small excess of nitrogen over growth requirements. At the time indicated the inflow was switched to a new medium containing the same level of nitrite and an equivalent amount of ammonium-N. Rates of uptake of nitrite (nitrosome activity) and of ammonia were calculated from changes in concentration within the culture vessel as in Fig. 2, and can be compared with the level of enzyme and of cellular intermediates. The division rate of the cells has been calculated from the rate of input of fresh medium necessary to maintain a constant cell density.
(a) ●, rate of nitrite uptake; ○, rate of ammonia uptake; - - - - -, rate of cell division; ———, level of enzyme/cell expressed on an arbitrary scale in order to match the steady-state enzyme level with the original rate of nitrite uptake; this enables rates per unit of enzyme in excess of the original rate to be easily recognized. The falloff in enzyme level is more than twice as fast as would occur if synthesis had stopped and cell growth had led to dilution of already existing enzyme; this indicates that appreciable destruction of enzyme must also occur. (b) △, total pool of amino acids; □, total nucleotide pool; ○, pool of free ammonia within the yeast cells; ●, level of nitrite in the culture medium.

already reduced the enzyme level to about 60% of that originally present, so that the rate of reduction/mole of enzyme was actually higher than that originally, in spite of the greatly increased concentrations of feedback inhibitors. However, this phase occurred at a time when the level of nitrite in the culture medium had risen to about twelve times the initial steady-state concentration, so it is possible that substrate-inhibitor competition for a *K* type allosteric enzyme (Monod *et al.*, 1965) could account for this apparent anomaly. Even in this experiment, where nitrite levels continued rising, it is clear that repression of the enzyme resulted in further reduction of nitrite uptake and eventually it did fall to zero; this suggests that at low substrate levels feedback inhibition may be the chief and immediate form of regulation but at high concentrations of substrate, at least when the enzyme shows *K* type kinetics, enzyme repression may be the dominant form of control.

MECHANISMS IN HIGHER PLANTS

Under certain conditions higher plants are, like yeast, exposed to a wide range of nitrogen compounds which they are able to assimilate. Thus, in the seedling, the breakdown of seed proteins provides amino acids and amides which in the growing tissues undergo extensive resynthesis and interconversion. With increasing maturity most seedlings switch to inorganic nitrogen assimilation but it is clear that the ability to utilize organic nitrogen can be retained by mature cells, since tissue cultures can be made to grow on either nitrate or amino acids. The potential of higher plants to use a variety of nitrogen sources can be readily investigated with the duckweed, *Lemna minor*, since it is particularly easy to culture this plant under axenic conditions. Unlike food yeast, this organism, although capable of sustained growth on organic sources of nitrogen, is unable to grow on these at rates comparable with those exhibited on nitrate or ammonia and the process of adaption to such sources is prolonged (Young, 1967). Nitrite here is used less readily than either nitrate or ammonia, plants growing more slowly and with a lower nitrogen content/frond. When supplied in combination with nitrate, amino acids bring about a real but limited sparing of nitrate assimilation; with nitrite there is no apparent sparing, but instead the plants are then able to increase their nitrogen content. These differences in physiological behaviour, as compared with food yeast, are indicative of fundamental differences in both metabolic and regulatory patterns but it is clear nevertheless that a flexible control system is still operative.

One of the complicating factors in the metabolism of higher plants is the fact that a particular catalytic function may be carried out by more than one enzyme. These distinct enzymes often have markedly different regulatory properties and until more is known about their localization within the plant, especially in relation to the variable disposition of metabolite pools, it is difficult to expose the direct relations which must exist between the concentration of intermediates and the level and activity of individual enzymes. Glutamic dehydrogenase offers one example of this. There is, in higher plants, an NADPH specific enzyme which is almost identical with that from yeast, being activated by α-oxoglutarate and inhibited by amino acids. This enzyme appears to be widely distributed in the roots of plants but we have been unable to detect it in plant leaves; in these a quite different glutamic dehydrogenase occurs, one that is NADH specific, and it appears probable that this enzyme is responsible for glutamate synthesis in these organs since it can be shown to be subject to repression and inhibition by exogenous amino acids (Stewart, unpublished). If so, this is in direct contrast to the situation in yeast, where the NADH linked enzyme has a catabolic function, serving to provide, by amino acid breakdown, the free ammonia needed for the synthesis of glutamine and carbamyl phosphate; it is formed in cells only when the free ammonia level falls sufficiently (see, for example, Fig. 8) and is induced by amino acids. It seems to us that in higher plants this division of function of the two pyridine nucleotides between catabolic (NAD) and anabolic (NADP) reactions no longer holds and instead their roles may depend upon the source of energy for synthesis, enzymes utilizing directly photosynthetic energy being NAD linked.

Circumstantial support for this view comes from nitrate reduction systems. In many higher plants nitrate reductase occurs as two distinct enzymes; the NADH specific enzyme is normally predominant, especially in leaves, but when *Lemna* plants are grown with an exogenous source of sucrose the level of the NADPH specific enzyme increases more than twentyfold, and there are also considerable increases of the NADPH generating enzymes, isocitrate dehydrogenase and glucose-6-phosphate dehydrogenase (Young, 1967). These changes are associated with a considerable decrease in carbon dioxide fixation in photosynthesis. The regulatory behaviour of the two enzymes is quite different (Fig. 13). The NADPH linked reductase is, like that of yeast, inhibited by amino acids, nucleotides and carbamyl phosphate, even at the lower concentrations at which these occur in higher plant cells. The NADH enzyme is affected only by amino acids.

Earlier investigations offer little support for the existence of a nitrosome in higher plants, since there are indications that the synthesis of the nitrate and nitrite reductases are independently regulated (Ingle *et al.*, 1966). However, it is likely that a multi-enzyme complex is responsible for the reduction of nitrite, since preparations are capable of forming ammonia directly from nitrite (Hewitt *et al.*, 1968). Decisive proof that an obligate association (within a nitrosome) of all the enzymes concerned with the reduction of nitrate to ammonia does not occur in higher plants has come from several experiments. By appropriate adjustment of inducer and repressor levels it is possible, in *Lemna*, to obtain synthesis of nitrite reductase independent of nitrate reductase, and moreover, during induction experiments, the ratio of these two activities changes considerably (Fig. 14). The nitrite reductase also differs from nitrate reductase in that it is not sensitive to feedback inhibition by amino acids or nucleotides (Fig. 13). This

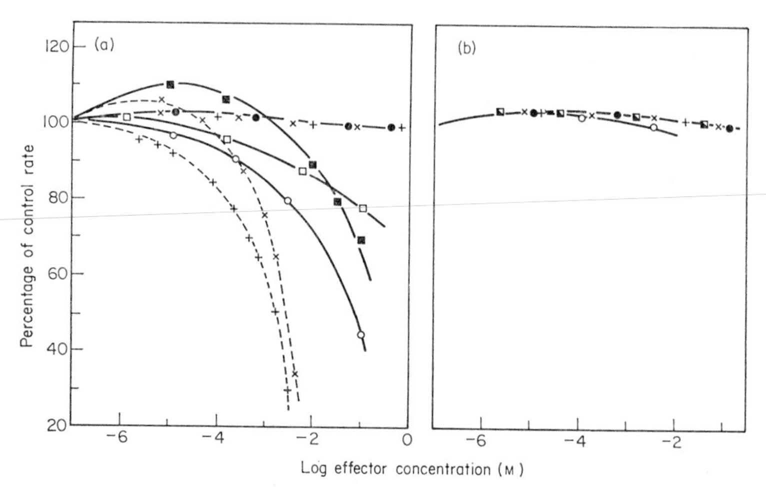

FIG. 13. The effects of soluble nitrogenous intermediates on the nitrate and nitrite reductases of the higher plant, *Lemna minor* (Stewart, Young and Sims, unpublished). Rates are shown as a percentage of that in the absence of allosteric effector.
(a) Effects on the two nitrate reductases; NADH enzyme ——— and NADPH enzyme - - - - -. (b) Effects on the nitrite reductase. ○, ammonia; ●, mixture of nucleotides (as in Fig. 11); ×, carbamyl phosphate; +, pyridoxamine phosphate; ◪, mixture of amino acids (as in Fig. 11); ■, aspartic acid; □, glutamic acid. ;

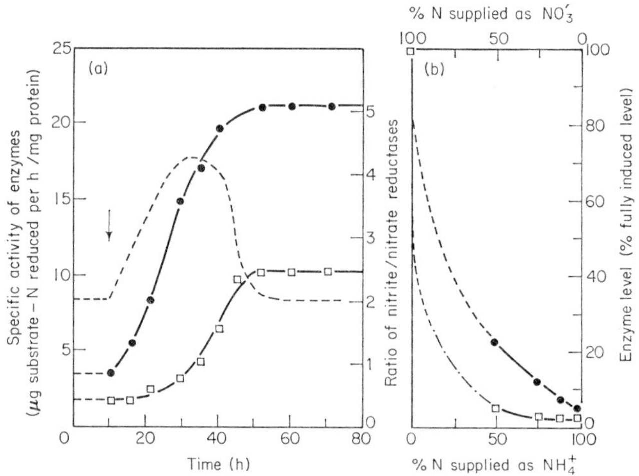

FIG. 14. Uncoordinate induction and repression of nitrate and nitrite reductases in *Lemna minor* (Stewart and Sims, unpublished).
(a) Induction with nitrate. Plants growing on ammonia medium were transferred to a nitrate medium at the point indicated. (b) Repression with ammonia. Plants were grown on media containing varying proportions of their nitrogen in the form of ammonia and of nitrate; the levels of the enzymes in plants after several generations on these media are expressed as a percentage of those in plants grown on nitrate alone.
□, nitrate (NADH) reductase; ●, nitrite reductase; - - - - -, ratio of nitrite to nitrate reductase.

difference in feedback behaviour serves to explain the results of the sparing experiments on *Lemna* described earlier and also offers a biochemical explanation for the toxicity to higher plants of high concentrations of nitrite (Harley, 1965).

A comparison of the regulatory properties of the corresponding enzymes from yeasts and higher plants gives some understanding of the way in which modification of control mechanisms may be involved in environmental adaptation. Thus, when nitrite reductase is considered, it is clear that the yeast, by virtue of its fully integrated nitrosome, is capable of regulating the metabolism of nitrite much more efficiently than is the higher plant, where the nitrite reductase is almost unregulated. However the low availability of nitrite in most soils may mean that there is little selective pressure for the retention of a regulated reductase in higher plants, so the lack revealed in laboratory experiments may have little biological consequence under natural conditions. A further point of difference in the assimilatory systems is the fact that the yeast enzymes are appreciably less sensitive to feedback inhibitors than are those from higher plants. In the yeast the regulatory systems show a sharp cut off at the rather high pool levels normally maintained in the cell, and this enables the enzymes to operate in the presence of the large concentrations of exogenous substrates which the yeast is likely to meet. Where the higher plant is concerned, studies on isolated enzymes are perhaps less informative, since the possibility exists that the greater structural complexity may lead to an isolation of functional components from the external environment. However, *Lemna*, which is probably less protected from its environment than many plants, is capable of dealing with low concentrations of a balanced mixture of amino acids, but shows symptoms of severe toxicity at higher concentrations which would be readily exploited by yeasts. Such levels are not a feature of the natural environment of most higher plants and the low availability of any combined nitrogen means that the amino acid and nucleotide pools are many times smaller than those in yeasts. These cellular intermediates are still able to act in a regulatory rôle because the response characteristics of the assimilatory enzymes have been correspondingly adjusted; this adjustment, however, has been at the cost of reduced tolerance of the presence of amino acids in the external environment.

ACKNOWLEDGEMENTS

We wish to thank the following colleagues who have contributed to this work through their individual participation in many of the experiments described here: Rowena B. K. Jones, P. E. Stanley, G. R. Stewart, B. Wurzberger and M. Young. We also gratefully acknowledge financial assistance from the Science Research Council in the form of research studentships for A. H. Bussey and many of the above, and a personal research grant for B. F. Folkes and A. P. Sims.

REFERENCES

Buchanan, J. M. and Hartman, S. C. (1959). *Adv. Enzymol.* **21**, 199.
Bussey, A. H. (1966). Ph.D. Thesis, University of Bristol.
Gilbert, D. A. (1959). Ph.D. Thesis, University of Bristol.
Harley, A. M. (1965). D.Phil. Thesis, University of Oxford.
Hewitt, E. J., Hucklesby, D. P. and Betts, G. F. (1968). See p. 47. of this book.
Ingle, J., Joy, K. W. and Hageman, R. H. (1966). *Biochem. J.* **100**, 577.
Kurland, C. G. and Maaløe, O. (1962). *J. molec. Biol.* **4**, 193.
Monod, J., Changeux, J.-P. and Jacob, F. (1963). *J. molec. Biol.* **6**, 306.
Monod, J., Wyman, J. and Changeux, J.-P. (1965). *J. molec. Biol.* **12**, 88.
Neidhardt, F. C. (1966). *Bact. Rev.* **30**, 701.
Reichard, P. (1959). *Adv. Enzymol.* **21**, 263.
Sims, A. P. (1958). Ph.D. Thesis, University of Bristol.
Sims, A. P. and Folkes, B. F. (1964). *Proc. R. Soc. B159*, 479.
Stadtman, E. R. (1966). *Adv. Enzymol.* **28**, 41.
Stanley, P. E. (1964). Ph.D. Thesis, University of Bristol.
Young, M. (1967). B.Sc. Thesis, University of Bristol.

Discussion on Section 1

In reply to Dr O. T. G. Jones' enquiry about experimental evidence that metals of metalloporphyrins can form complexes with nitrogen, such as had been described for other metallo-organic complexes, Professor Chatt said: "Interaction between nitrogen gas and metalloporphyrins had been reported, but before nitrogen complexes were discovered no one knew what to look for as a diagnostic test. Now we know that a strong band in the infrared spectrum in the region 2050–2170 cm^{-1} occurs in all nitrogen complexes. Such a band may be caused by hydrogen attached to the metal but usually hydrogen gives a band at lower frequency. We have found a few complexes which develop absorptions in this region on passing nitrogen through their solutions, but no product containing the band could be isolated and characterized. It does appear that unstable nitrogen complexes are more common than supposed, perhaps the metalloporphyrins form such complexes; we have done no work on them ourselves".

In commenting on recurring reports that spectroscopic shifts can be detected when nitrogen is added to suitably reduced leghaemoglobin, he did not wish to discount these, but they may be based on too little evidence.

On the prospects for commercial application of the newly discovered complexes of nitrogen to the production of ammonia Professor Chatt commented: "Everyone is waiting for the answer to this question. Those who discovered the complexes are all working in different fields of chemistry and discovered the complexes by accident. So far we know very little about the reactions of the complexes. Everyone is looking for new complexes. We know nothing other than I mentioned in my lecture, that is relevant to nitrogen fixation. The complex $[Co(N_2)(PPh_3)_3]$ is stated to give off all of its nitrogen as gas at just over 100°C; $[CoH(N_2)(PPh_3)_3]$ gives off all of its nitrogen and hydridic hydrogen as gas at 100–150°C. It would be interesting to see whether hydrolysis produces any hydrazine. If reduction of this complex to ammonia and the original hydride $[CoH_3(PPh_3)_3]$ could be effected then the hydride would pick more nitrogen for reduction and a possible process for low pressure reduction of nitrogen to ammonia would be born. Whether this would revolutionize the synthetic ammonia industry is another matter. I think that it gives a clue as to how biological fixation occurs, but the biological system would be much more complicated".

In reply to Dr M. G. Yates' comment that studies on purified nitrogen-fixing systems from free living organisms suggest a non-haem-iron association with the fixing site, Professor Chatt indicated that Bulen (1966) had subdivided purified nitrogenase from *Azotobacter vinelandii* into two fractions which do not fix nitrogen independently, but do so when added together. Iron is associated with both these fractions but not as a haem compound. Similar results have been obtained by Mortenson (1966) with purified nitrogenase from *Clostridium pasteurianum*. If one can infer a similar "active site" of nitrogenase in all biological systems, these results suggest that haem compounds do not function in the active site.

Dr E. J. Hewitt, in commenting on the system which Turchin *et al.* have described, and which Professor Bond had mentioned, said "They made no distinction between legumes and non-legumes. Considering legumes first of all: in the nodules of the lupin we might obtain an enzyme for fixation if the endophyte is present. There might be a compound something like ferredoxin in the extracts because Sanderson and Cocking in their work on the reduction of nitrite by tomato tissues suggested there was a compound something like ferredoxin in the non-green parts of the plants which could react with photochemical systems. This has not been isolated but nonetheless the suggestion is very important. If this is correct, there could be a compound of this sort present in the non-green parts of leguminous plants, including the root nodules. The one thing which Turchin *et al.* have done, which I think no-one else has done in this context, is to add ATP with an ATP generating system; this puts their work on a par with all the other work on *Azotobacter*, *Clostridium* and so on, in having an ATP generating system with ferredoxin or some other non-haem iron protein in a nitrogen fixing system. It might also be expected that from the nodules some type of B_{12} compound would have been extracted and Professor Bond's suggestion that Co in B_{12} or other organo-compounds may be reacting with nitrogen is an important one, particularly in this context. When considering cereals, however, we do not think that higher plants contain a B_{12} type of compound capable of supporting the growth of any "assaying" micro-organism though we know perfectly well from Nicholas's work that even a fungus grown with Co will produce compounds which are related to B_{12}. Plants also may produce some type of cobalamine compound or cobalt complex which is present in the leaves possibly, and this complex could be involved in nitrogen fixation when ATP and ferredoxin are also present. But why does this system not function more obviously if present?"

Dr A. J. Willis referring to the work of Hassouna and Wareing, said that in the Ph.D. thesis of M. G. Hassouna ("Soil Micro-organisms and Nutrient Relations in Plant Colonization") *Azotobacter* rather than a mycorrhizal fungus is thought to be an important source of nitrogen for marram (*Ammophila arenaria*). It is shown that large populations of these bacteria are often closely associated with the roots of the grass.

Professor Bond replied that his only source of information had been Hassouna and Wareing's published article, and there the mycorrhizal association was mentioned, among other possibilities, as a likely source of nitrogen.

Arising from Dr Rosalie Cox's paper, Dr Hewitt enquired: "Have you added NADP to your *Anabaena* cultures in the same way as you have added pyruvate. We find, and Shin and Oda find exactly the same thing, that NADP totally inhibits the reduction of nitrite in a photo-chemical system in which the two are mixed together and no nitrite is reduced until all the NADP is used up. This can be related to the flavoprotein reducing NADP by ferredoxin, for which the K_m is about 3×10^{-7} M as compared to the K_m of nitrite reductase/ferredoxin which is 10^{-5} M. If we suppose that ferredoxin is the final currency for electron donation in nitrogen fixation in photosynthetic organisms, we expect the same sort of situation, that NADP might be a powerful inhibitor and might bring the reaction to a standstill and that the balance between the extent to which endogenous NADP was in a reduced form would determine the ratio of oxygen evolution to nitrogen fixation.

Dr Cox replied that she had not tested the effect of NADP on nitrogen fixation in *Anabaena* cultures but she doubted very much whether this substance would enter the cells.

After Dr B. J. Miflin's paper "Nitrate reducing enzymes in barley" Dr E. C. Cocking suggested that in considering aspects of nitrate assimilation by roots it was important to take into account the possible effects of contaminating micro-organisms on the results obtained and also of different extraction conditions, including, for instance, the effects of sulphydryl compounds and polyvinylpyrrolidone on the level and distribution of the enzymes involved between the cytoplasm and organelles such as mitochondria. How far had Dr Miflin assessed these possible factors in his work?

The author explained that roots were routinely grown in open cultures in apparatus which had been washed with detergent and sterilized with hypochlorite, and the solutions were changed daily. The roots were washed before doing experiments. Barley roots had been grown aseptically and these contained active nitrate and nitrite reductase, but the experiments described in the paper referred to roots grown in open culture. The activity of the mitochondria is affected by the method of extraction. The use of a Waring blender which has a harsh action and which is difficult to keep cold leads to largely inactive mitochondria measured in terms of respiratory as well as nitrate and nitrite reductase activity. He had obtained the best preparations by gentle grinding in a well chilled pestle and mortar without using sand. As a routine 10^{-3} M glutathione was included in the extraction medium. Polyvinylpyrrolidone had also been used but did not appear to affect greatly the distribution of activities between the supernatant and mitochondrial fractions.

Dr Miflin agreed with Dr Hewitt's suggestions that there was every reason to look for a one-electron donor in root systems; there was no doubt from the work of Dr Miflin and others that nitrate and nitrite can be reduced by a complex of enzyme systems when viologen dyes are present. In practically every case which has been studied (except the HR2 system described in his own paper) viologens and ferrodoxins are interchangeable with varying degrees of efficiency and, in plants, ferredoxins are one-electron donors. A search should be made for as-yet-unidentified one-electron donors in roots as these are probably the key to the second stage of nitrate reduction in roots, namely, nitrite to ammonia.

On the paper by Drs B. F. Folkes and A. P. Sims, Dr J. S. Pate enquired whether one of the two nitrate reductase systems recovered from *Lemna minor* might be localized in chloroplasts, and the other in the cell cytoplasm. The authors considered that one was chloroplastic and the other from a source in the roots.

Dr Pate also pointed out that evidence had been provided of a delicate relationship between rates of synthesis and sizes of pools of glutamate and glutamine in yeast cells. Could this relationship be altered by changing the level of free ammonia or carbon source in the medium?

Dr Sims explained that they had tried to maintain yeast in a steady state over a fairly wide range of ammonia concentrations and this certainly had affected the steady state concentration of glutamine on the pool.

Dr Miflin asked: "Have you any evidence that you only have one enzyme system from NADPH to nitrate? Losada and his co-workers (*Biochim. biophys. Acta* **109**, 79; and **128**, 202) have separated out a nitrate reductase which will accept electrons from reduced FMN or NADH or from an NADPH diaphorase with FMN. Combination of the two enzymes leads to reduction of nitrate by NADPH."

Dr Sims replied: "Dr Folkes and myself were trying to suggest that we should know how many enzymes are working with NADH or NADPH or what the source of energy is. For example, you can get amination mutants of *Neurospora* which are in fact deficient in NADP activity. Under these unique conditions you can still get growth; there is sufficient build-up of ammonia in the cell to enable the NAD enzyme to work in reverse. We are trying to ascertain the exact synthetic pathways in order to be able to appreciate the possible interactions of the substrates and end products of the enzyme system. We have attempted a generalization which may not be complete, but it is essential to make some form of generalization in order to orientate further study."

Dr Miflin further referred to the statement that the levels of three NADPH linked enzymes are all increased in the presence of sucrose. Was there any evidence that the compounds which repress NADPH nitrate reductase also repress the other two enzymes.

Dr Sims explained that it had been possible to influence the ratio of NADPH enzyme to NADH enzyme. With regard to nitrate reductase in *Lemna* it is possible to reduce nitrate to nitrite using either NADH or NADPH as electron donors but this is not to say they are involved in the process; the maximum activity is still with ferredoxin or green plant extracts.

Professor D. D. Davies asked: "If in your laboratory the NADP specific glutamic dehydrogenase you are measuring in higher plants can only be detected in older roots and not in leaves, whilst in my laboratory we cannot detect this enzyme in seedlings, is it possible that you are picking up bacterial contamination in the roots?"

Dr Sims replied that he did not think so because Mr M. Young had been growing *Lemna* in sterile conditions, and where sucrose was given in conditions under which NADP reductase systems go up enormously, this high level of glutamic dehydrogenase in the plants was obtained. In this case it could not be due to bacterial contamination because they were sterile plants. Dr Pate would refer to a similar effect, i.e. the NADP systems may somehow be associated with the carbohydrate supply to the roots. A lot may depend on the nitrate/carbohydrate regime of the organism as to just how much of the reduction is due to the NADP and how much to the NAD systems.

Professor E. W. Yemm asked whether only the NAD system was found in leaves. Was there no evidence of the NADP system there as well, but Dr Sims explained there was some, but not very much, in the roots.

Dr Hewitt thought one would have found NADP system in chloroplast grana and chloroplast + leaves where NADP is the direct electron acceptor and transhydrogenase is required in any case to activate NAD. He commented that Ritenour *et al.* (1967) found that NAD specific GDH (according to their reference to Bulen) is not associated with chloroplasts; it did not appear that they had looked for any NADP specific GDH activity in chloroplasts but some leakage would have been expected in Bulen's method of extraction. Even so, Bulen maintained that the system was NAD specific.

Section 2

Intermediate Metabolism of Amino Acids and Relationships with Mineral Nutrition and Protein Synthesis

Section 2a

The Metabolism of Amino Acids in Plants

DAVID D. DAVIES

School of Biological Sciences
University of East Anglia, Norwich, England

INTRODUCTION

The publication of Alton Meister's two volumes on the "Biochemistry of Amino Acids" bears witness to the rapid development of knowledge in this area, but alas reminds us that the advance of plant biochemistry is slow.

In the absence of experimental data it is necessary to assume the essential similarity of intermediary metabolism between bacteria, plants and animals. In general the assumption is valid for the main metabolic pathways, e.g. the Embden–Meyerhof pathway and the Krebs cycle. However we have sufficient evidence to suggest caution in the application of the assumption to amino acid metabolism. Admittedly there has been no experimental support for the suggestion (Smith *et al.*, 1961) that photosynthetic tissue contains specific enzymes for the direct reductive amination of phosphoenolpyruvate and oxaloacetate:

$$\text{phosphoenolpyruvate} \xrightarrow[\text{NH}_4{}^+]{\text{NADPH}} \text{alanine} + \text{NADP} + \text{P}_i$$

$$\text{oxaloacetate} \xrightarrow[\text{NH}_4{}^+]{\text{NADPH}} \text{aspartate} + \text{NADP}$$

On the other hand, a number of other differences have been recorded. For example, the proposal of Vogel (1959) that lysine biosynthesis in most fungi proceeds *via* aminoadipic acid and in plants and bacteria *via* diaminopimelic acid is now well supported. Thus when diaminopimelic acid-1, 7-^{14}C is fed to wheat plants, lysine-1-^{14}C is formed, but aminoadipic acid-6-^{14}C does not produce lysine (Finlayson and McConnell, 1960; Nigam and McConnell, 1963). Furthermore, the terminal enzyme of the sequence—diaminopimelic decarboxylase has been shown to be present in plants (Shimura and Vogel, 1961) but is absent from fungi. Both pathways are outlined in Fig. 1.

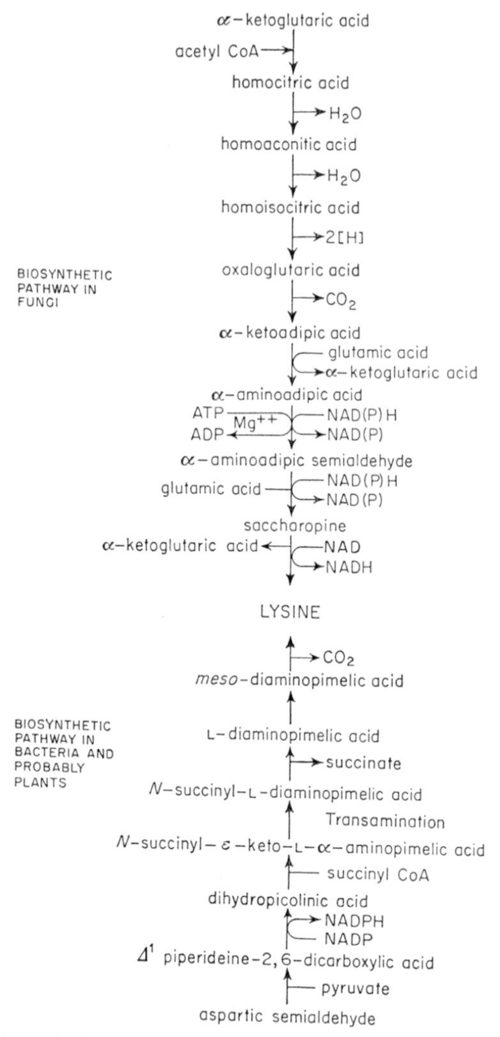

FIG. 1. Biosynthesis of lysine *via* the α-aminoadipic and diaminopimelic acid routes.

The assay of saccharopine dehydrogenase has been suggested as a method of confirming the absence of the aminoadipic acid pathway in plants (Vaughan and Broquist, 1965).

A second example is to be found in the biosynthesis of the aromatic amino acids phenylalanine and tyrosine. There is good evidence that in plants and bacteria they are formed by the reactions shown in Fig. 2.

Animals cannot synthesize aromatic amino acids, but can hydroxylate phenylalanine to give tyrosine. The cofactor for phenylalanine hydroxylase

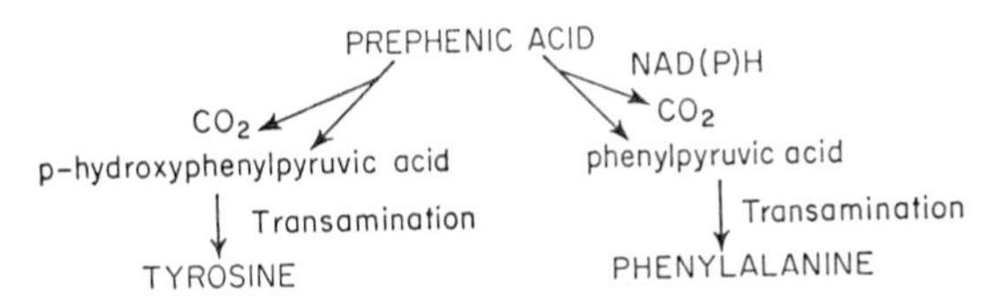

FIG. 2. Biosynthesis of tyrosine and phenylalanine in bacteria and plants.

has been found to be a dihydrobiopterin (Kaufman, 1963, 1964). Bacteria do not normally hydroxylate phenylalanine, but an adaptive pathway which is similar to the animal system, can be induced by growth on phenylalanine (Guroff and Ito, 1965). The situation in plants is still somewhat confused. Thus feeding experiments have indicated that in plant tissue culture, phenylalanine and tyrosine do not contribute carbon to each other (Dougall, 1965). Other feeding experiments indicate that *Salvia splendens*, *Triticum vulgare* and *Fagopyrum tataricum* can convert phenylalanine to tyrosine by direct hydroxylation (McCalla and Neish, 1959; Gamborg and Neish, 1959). Nair and Vining (1965) have purified a phenyalanine hydroxylase from spinach leaves. The enzyme shows maximum activity with tetrahydrofolate (THFA) and NADH, and spinach leaves contain a cofactor which can substitute for THFA. These results suggest that if plants are kept in air enriched with $^{18}O_2$, the tyrosine should show ^{18}O enrichment. Experiments with etiolated soya bean and maize seedlings have shown virtually zero enrichment of ^{18}O in tyrosine (Fritz and Aman, 1966), but results for spinach are not yet published.

The third example is the pathway of transsulphuration involving cystathionine as an intermediate. In mammals, the reactions involved are:

$$\text{homocysteine} + \text{serine} \rightarrow \text{cystathionine} \rightarrow \text{cysteine} + \alpha\text{-ketobutyrate} + NH_3$$

In bacteria, the reactions involved are:

$$\text{cysteine} + O\text{-succinylhomoserine} \rightarrow \text{cystathionine} + \text{succinate}$$
$$\downarrow$$
$$\text{homocysteine} + \text{pyruvate} + NH_3$$

In fungi, both transsulphurations occur though it is probable that the source of activated homoserine is *O*-acetylhomoserine rather than *O*-succinylhomoserine (Nagai and Flavin, 1966).

In plants, the reactions involved are similar to those of bacteria except that in spinach, *O*-succinylhomoserine and *O*-acetylhomoserine are equally effective as precursors of cystathionine (Giovanelli and Mudd, 1966).

It should be noted that *O*-acetylhomoserine has been isolated from peas (Grobbelaar and Steward, 1958).

This introduction is becoming unduly long and I therefore propose to present a more personal interest in the amino acid metabolism of plants.

THE BIOSYNTHESIS OF GLYCINE, SERINE AND METHIONINE

A number of reactions involving the amino acids glycine, serine and methionine are illustrated in Fig. 3.

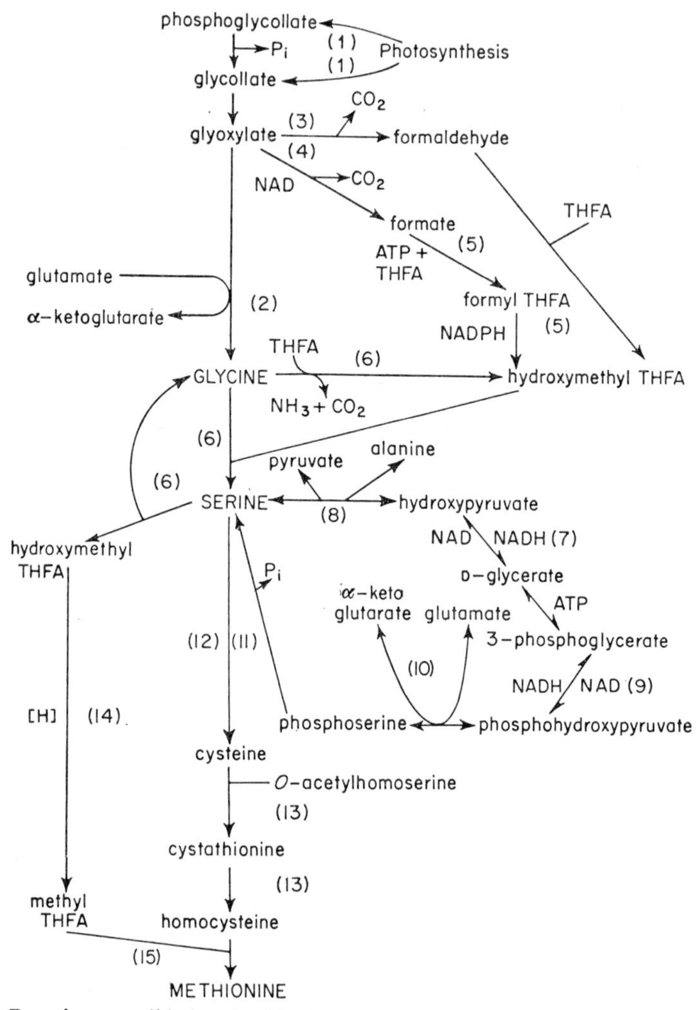

FIG. 3. Reactions possibly involved in the metabolism of glycine, serine and methionine. (The numbers refer to reactions discussed in the text)

The following points may be noted:

(1) The formation of glycollate during photosynthesis is a major unsolved problem of plant biochemistry. Zelitch (1966), using tobacco leaves, found that after 2 min of photosynthesis in $^{14}CO_2$, glycollate was uniformly labelled and had a specific activity higher than the carboxyl group of 3-phosphoglycerate. Hess and Tolbert (1966a), also using tobacco leaves, found that the specific activity of glycollate was always below that of the carboxyl of 3-phosphoglycerate and also below that at carbons 2 and 3. They argue that 3-phosphoglycerate could be the precursor of glycollate, but their contention that glycollate is not the precursor of 3-phosphoglycerate seems invalid. Serine was equally labelled in carbons 1 and 2 when phosphoglycerate and glycerate were 5-fold more carboxyl labelled than carbons 2 and 3. In tobacco, serine thus appears to be formed from uniformly labelled glycollate. In *Chlorella*, serine has the same distribution of ^{14}C as 3-phosphoglycerate. Since algae do not have glycollic acid oxidase, serine was probably formed from glycerate or phosphoglycerate (Hess and Tolbert, 1966b).

(2) The transamination reaction between glyoxylate and amino acids has been studied by Cossins and Sinha (1965). They observed transamination with a number of amino acids. Unfortunately they did not include controls for non-enzymic reactions which have been shown to occur between glutamate and glyoxylate at pH 7·4 (Nakada and Weinhouse, 1953). There is also some uncertainty about the reversibility of this transamination. Nakada (1964) has reported that the reaction between glyoxylate and glutamate is irreversible. Cossins and Sinha on the other hand reported reversibility using oxaloacetate and pyruvate as the amino acceptors from glycine.

(3) The decarboxylation of glyoxylic acid was studied by Dr J. R. Corbett and myself using enzymes isolated from various plants. Our results are consistent with the view that glyoxylate is a substrate for pyruvic decarboxylase. The kinetics of this decarboxylation show a sigmoid relationship in the plot of velocity against glyoxylate concentration. This curve is frequently associated with allosteric enzymes—in this case we believe that it indicates a second order reaction. The rate determining step in the overall reaction is the dissociation of formaldehyde from the enzyme–hydroxymethyl–thiamine pyrophosphate complex. It is assumed that this complex reacts with a second molecule of glyoxylic acid. In support of this mechanism we have been able to show that the following reactions are catalyzed by pyruvic decarboxylase:

$$\text{glyoxylate} + \text{acetaldehyde} \rightleftharpoons \text{lactaldehyde} + CO_2$$

$$\text{glyoxylate} + \text{glycolaldehyde} \rightleftharpoons \text{glyceraldehyde} + CO_2$$

(4) We have isolated pyruvic oxidase from mitochondria and shown that glyoxylate is a substrate for, and also a competitive inhibitor of this enzyme. Possibly this inhibition of pyruvic oxidation could be associated with the light inhibition of respiration.

(5) Formyltetrahydrofolate synthetase has been purified from plants and shown to require high concentrations of K^+(0·2–0·3M) for maximum activity (Hiatt, 1965). The formation of hydroxymethyl THFA from formyl THFA has been demonstrated with preparations from peas (Wilkinson and Davies, 1960).

(6) Serine aldolase or serine hydroxymethyl transferase was first demonstrated in plants by Wilkinson and Davies (1958). More recently these results have been confirmed by Cossins and Sinha (1966) and evidence was obtained with cell-free extracts that plants can catalyze the overall reaction:

$$2 \text{ glycine} \rightarrow \text{serine} + CO_2 + NH_3$$

The individual reactions are:

$$\text{glycine} + \text{THFA} \xrightarrow{H_2O} \text{hydroxymethyl THFA} + NH_3 + CO_2$$

$$\text{glycine} + \text{hydroxymethyl THFA} \rightleftharpoons \text{serine} + \text{THFA}$$

(7) Whereas lactic dehydrogenase can convert L-glycerate to hydroxypyruvate, in plants D-glycerate dehydrogenase is a widely distributed specific dehydrogenase (Stafford *et al.*, 1954).

(8) A transaminase active with hydroxypyruvate and alanine or glycine has been demonstrated in plants—the enzyme is particularly active in leaves (Willis and Sallach, 1964).

(9) Phosphoglycerate dehydrogenase has been purified from peas by C. Slaughter in my laboratory. The enzyme is more active than glycerate dehydrogenase in extracts of etiolated tissue.

(10) C. Slaughter has also obtained evidence for a transamination between phosphohydroxypyruvate and glutamate in pea extracts.

(11) Phosphoserine is a substrate for the general acid phosphatase of peas but peas also contain a phosphatase with a very sharp pH optimum at pH 7·3 and relatively high activity with phosphoserine.

(12) In animals, cysteine is made from serine by transsulphuration from homocysteine. In yeast, cysteine is made by the reaction:

$$\text{serine} + H_2S \rightarrow \text{cysteine} + H_2O$$

In bacteria, cysteine is formed by the reaction:

$$O\text{-acetylserine} + H_2S \rightarrow \text{cysteine} + \text{acetate} + H_2O$$

The mechanism in plants is unknown.

(13) The formation of homocysteine from cysteine and *O*-acetylhomoserine has been discussed in the introduction.

(14) The reduction of hydroxymethyl THFA to methyl THFA has not been demonstrated in plants. In animals, it requires NADPH or NADH and a flavoprotein.

(15) The biosynthesis of methionine in animals involves a cobamide enzyme. Laties and Hoelle (1967) have observed the conversion of propionate to succinate in potato slices and the labelling data is consistent with the reaction sequence:

$$\text{propionate} \xrightarrow{CO_2} \text{methylmalonyl CoA} \rightarrow \text{succinate}$$

The conversion of methylmalonyl CoA to succinate in bacteria has a vitamin B_{12} dependence but the mechanism of this isomerization in plants is obscure. The bulk of evidence is against the presence of vitamin B_{12} in plants but Graeser (1965) has presented evidence that vitamin B_{12} stimulates serine production in extracts of roots.

As far as methionine biosynthesis is concerned, a likely pathway is *via* a triglutamate form of methyltetrahydrofolic acid rather than the usual monoglutamate form.

The failure to demonstrate the reduction of hydroxymethyl groups to methyl groups using cell-free preparations, necessitates the use of feeding experiments. The results of feeding formaldehyde-^{14}C to washed discs of swedes is shown in Tables I and II.

Formaldehyde reacts with amino groups (formol titration), hence to compare the effects of amino acids on the metabolism of formaldehyde, it is necessary to use the D-forms as controls. The results shown in Table I

TABLE I

Effect of various compounds on the formation of $^{14}CO_2$ from formaldehyde-^{14}C. Formaldehyde-^{14}C (0·4 μ moles) incubated in 2ml potassium phosphate buffer, (0·01M,pH 6·0) with 16 swede discs (washed 48 hours) and 5 μ moles of the compounds listed in the table. Temp. 30°C.

Additions	HCHO uptake c.p.m.	O_2 uptake μl	$^{14}CO_2$ c.p.m.	$^{14}CO_2$ c.p.m./μl	$^{14}CO_2$ c.p.m./μl/10^4c.p.m. HCHO uptake
Nil	10,500	178	5,300	3	3
D-Methionine	12,700	169	5,900	3·5	3
L-Methionine	13,200	100	850	0·8	0·6
D-Serine	13,100	182	5,200	2·9	2·2
L-Serine	13,500	184	1,100	0·6	0·5

indicate that amino acids increase the uptake of formaldehyde and that the D-forms of methionine and serine have little effect on the specific activity of $^{14}CO_2$ formed from formaldehyde-^{14}C. On the other hand, L-methionine and L-serine markedly reduce the specific activity suggesting a possible equilibration between formaldehyde, the hydroxymethyl group of serine and the methyl group of methionine.

This explanation was tested by isolating methionine, serine and methylcysteine at the end of the incubation. The amino acids were isolated by filter-paper chromatography, eluted and counted. The results are presented in Table II.

TABLE II

Effects of methionine, serine and methylcysteine on the incorporation of formaldehyde-^{14}C into various components of swede discs. Conditions as for Table I except that quantities were doubled.

Compounds supplied, uptake and recovery of ^{14}C	Methionine		Serine		DL-Methyl cysteine
	D	L	D	L	
Formaldehyde uptake (c.p.m.)	107,000	112,500	111,000	107,000	102,000
Total counts in compounds soluble in 80% ethanol	33,700	39,500	26,300	37,800	29,700
Counts in serine	8,500	1,600	7,000	12,100	5,100
Counts in methionine	800	27,900	500	500	800
Counts in methylcysteine	—	—	—	—	2,000

The incorporation of radioactivity from formaldehyde-^{14}C into methionine and methylcysteine was slight except when these compounds were supplied simultaneously with the formaldehyde-^{14}C.

Methylcysteine has been shown to be formed in yeast by the condensation of methylmercaptan and serine and a similar mechanism has been suggested in garlic (Sugii *et al.*, 1963). The data for swedes suggest that methyl-*S*-cysteine is formed by the methylation of cysteine. This explanation is confirmed by the recent finding that the synthesis of methylcysteine in radish leaves involves the methylation of cysteine by methionine (Thompson and Gering, 1966).

Similarly, the incorporation of ^{14}C into methionine is consistent with the reduction of a hydroxymethyl group to a methyl group and subsequent transfer to homocysteine.

An interesting feature of these results is that superficially they appear to show that control of methionine synthesis by end-product inhibition is absent in swede discs.

CONTROL OF AMINO ACID BIOSYNTHESIS

The control of amino acid biosynthesis in bacteria is well documented (see Stadtman, 1966), but once again there is a great lack of information concerning plants. End-product inhibition, the first reaction specific to a particular sequence, has been demonstrated in a few cases. Thus Davies (1964) has shown that acetolactate synthesis in peas is inhibited by valine. Oaks (1965) has shown that the enzyme from maize condensing 2-keto-isovaleric acid and acetyl CoA, is specifically inhibited by leucine. Slaughter has found that serine inhibits 3-phosphoglycerate dehydrogenase of pea seedlings. The inhibition is non-competitive, cold labile and specific.

Attempts have been made to detect end-product inhibition by feeding experiments. For example, Oaks (1965) measured the effect of exogenous amino acids on the incorporation of acetate-2-^{14}C into amino acids. Her results are recalculated and presented in Table III.

TABLE III

Effect of exogenous amino acids on the incorporation of acetate-2-^{14}C in amino acids of maize root tips.

(Excised root tips were pretreated for 3 hours in a solution of salts containing 1% sucrose and a mixture of amino acids. At this time the roots were washed with water and fresh salts, sucrose and amino acids were added together with acetate-2-^{14}C for an additional 2 hours.

Amino acids (mg/2 ml)		c.p.m./20 tips	
		Sucrose	Sucrose and amino acids
Serine	0·347	2,755	2,830
Glycine	0·168	2,629	2,348
Threonine	0·198	13,760	4,319
Alanine	0·476	29,560	31,860
Valine	0·33	5,725	2,846
Leucine-Isoleucine	1·009 and 0·237	35,560	9,369
Lysine	0·088	7,090	2,140
Arginine	0·15	11,745	9,110
Proline	0·621	25,990	3,980
γ-Aminobutyric		48,100	48,400
Glutamic	—	32,924	34,165
Aspartic	—	5,875	5,482
Methionine	0·117	—	—
Histidine	0·164	—	—

The results indicate that the synthesis of threonine, valine, leucine, lysine, arginine and proline from acetate-2-^{14}C was inhibited by exogenous amino acids. The experiment fails to distinguish adequately between end-product (allosteric) inhibition and enzyme repression.

An experiment by Dougall (1965) indicated that exogenous lysine, threonine, isoleucine, methionine, arginine, valine, leucine, serine, glycine, histidine, phenylalanine and tyrosine all inhibited the incorporation of glucose-^{14}C into the very amino acid of the protein. Unfortunately the incorporation of ^{14}C into soluble amino acids was not measured and consequently the results could be explained by pool dilution and cannot therefore be taken as evidence for end-product inhibition.

Many aspects of the control of amino acid metabolism show good design—the subject is ideal for teleology and following this approach there would seem to be two obvious points of control: (*a*) the fixation of carbon dioxide and entry into the Krebs cycle, and (*b*) the entry of ammonia into organic combination.

The Krebs cycle cannot produce a net gain of carbon and consequently when ketoacids (α-ketoglutarate and oxaloacetate) are removed from the cycle to form amino acids, there must be a corresponding fixation of carbon dioxide into oxaloacetate or malate. It would therefore be expected that these processes are geared to one another and it should be interesting to detect the method of gearing.

Glutamic dehydrogenase represents the main portal for the entry of NH_4^+ into organic combination. Until recently there was evidence only for a NAD specific glutamic dehydrogenase in plants. Folkes and Sims (1968 and personal communication, 1965) however, have obtained evidence that plants also contain a NADP specific glutamic dehydrogenase. It will be interesting to see if this enzyme shows allosteric properties comparable to the NADP glutamic dehydrogenases of micro-organisms.

REFERENCES

Cossins, E. A. and Sinha, S. K. (1965). *Can. J. Biochem.* **43**, 495.
Cossins, E. A. and Sinha, S. K. (1966). *Biochem. J.* **101**, 542.
Davies, M. E. (1964). *Pl. Physiol.* **39**, 53.
Dougall, D. K. (1965). *Pl. Physiol.* **40**, 891.
Finlayson, A. J. and McConnell, W. B. (1960). *Biochim. biophys. Acta* **45**, 622.
Folkes, B. F. and Sims, A. P. (1968). *See* p. 91 of this book.
Fritz, G. and Aman, F. (1966). *Pl. Physiol.* **41**, suppl. xiv.
Gamborg, O. L. and Neish, A. C. (1959). *Can. J. Biochem.* **37**, 1277.
Giovanelli, J. and Mudd, S. H. (1966). *Biochem. biophys. Res. Commun.* **25**, 366.
Graeser, H. (1965). *Naturwissenschaften* **52**, 5.
Grobbelaar, N. and Steward, F. C. (1958). *Nature, Lond.* **182**, 1358.
Guroff, G. and Ito, T. (1965). *J. biol. Chem.* **240**, 1175.
Hess, J. L. and Tolbert, N. E. (1966a). *J. biol. Chem.* **241**, 5705.
Hess, J. L. and Tolbert, N. E. (1966b). *Pl. Physiol.* **41**, suppl. xxxix.
Hiatt, A. J. (1965). *Pl. Physiol.* **40**, 189.

Kaufman, S. (1963). *Proc. natn. Acad. Sci. U.S.A.* **50**, 1085.
Kaufman, S. (1964). *J. biol. Chem.* **239**, 332.
Laties, G. G. and Hoelle, C. (1967). *Phytochemistry* **6**, 49.
McCalla, D. R. and Neish, A. C. (1959). *Can. J. Biochem.* **37**, 531.
Nagai, S. and Flavin, M. (1966). *J. biol. Chem.* **241**, 3861.
Nair, P. M. and Vining, L. C. (1965). *Phytochemistry* **4**, 401.
Nakada, H. I. (1964). *J. biol. Chem.* **239**, 468.
Nakada, H. I. and Weinhouse, S. (1953). *J. biol. Chem.* **204**, 831.
Nigam, S. N. and McConnell, W. B. (1963). *Can. J. Biochem.* **41**, 1367.
Oaks, A. (1965). *Pl. Physiol.* **40**, 149.
Shimura, Y. and Vogel, H. J. (1961). *Fedn Proc.* **20**, 10.
Smith, D. C., Bassham, J. A. and Kirk, M. (1961). *Biochim. biophys. Acta* **48**, 299.
Stadtman, E. R. (1966). *Adv. Enzymol.* **28**, 41.
Stafford, H. A., Magaldi, A. and Viennesland, B. (1954). *J. biol. Chem.* **207**, 621.
Sugii, M., Nagasaura, S. and Suzuki, T. (1963). *Chem. pharm. Bull.* **11**, 135.
Thompson, J. F. and Gering, R. K. (1966). *Pl. Physiol.* **41**, 1301.
Vaughan, S. T. and Broquist, H. P. (1965). *Fedn Proc.* **24**, 218.
Vogel, H. J. (1959). *Proc. natn. Acad. Sci. U.S.A.* **45**, 1717.
Wilkinson, A. P. and Davies, D. D. (1958). *Nature, Lond.* **181**, 1070.
Wilkinson, A. P. and Davies, D. D. (1960). *J. exp. Bot.* **11**, 296.
Willis, J. E. and Sallach, H. J. (1964). *Phytochemistry* **2**, 23.
Zelitch, I. (1965). *J. biol. Chem.* **240**, 1869.

Section 2b

The Biosynthesis of Putrescine in Higher Plants and its Relation to Potassium Nutrition

T. A. SMITH

Agricultural Research Council Unit
Wye College, Wye, Ashford, Kent, England

The function of potassium in plants is still a subject for much speculation. Although large amounts are required for normal growth, in only a few cases has potassium been found to be involved in enzyme systems. The discovery that the diamine, putrescine, accumulates in potassium-deficient barley leaves was therefore of considerable interest (Richards and Coleman, 1952).

On the basis of studies on the biosynthesis of putrescine by bacteria, it was anticipated that in higher plants this amine would be derived from arginine. F. J. Richards and Mary Bryant (unpublished work) confirmed this in barley, by feeding isotopically labelled arginine to the leaves, when labelled putrescine was produced. Two routes by which arginine may be converted to putrescine in bacteria are shown in Fig. 1. These are: either, *via* ornithine which may be decarboxylated to putrescine (Gale, 1940); or, *via* agmatine, the decarboxylation product of arginine, which can be converted to putrescine, either directly with the formation of urea (Miyaki and Momiyama, 1956), or through the intermediate production of *N*-carbamyl putrescine (Linneweh, 1932; Møller, 1955).

In barley, labelled ornithine was found to be only slowly converted to putrescine (Coleman and Hegarty, 1957). The direct route from ornithine to putrescine was therefore not thought to be the major pathway of putrescine synthesis. However, it was subsequently demonstrated that agmatine accumulates, together with the putrescine, in potassium-deficient barley (Smith and Richards, 1962). From 200 g fresh weight of potassium-deficient barley leaves, 2 mg of agmatine was extracted which was characterized by its melting point and infra-red spectrum.

Figure 2 shows the agmatine content of the leaves of three varieties of barley, grown under normal and potassium-deficient conditions (Smith, 1963). The agmatine appears to reach a peak at the tenth week. The putrescine level was frequently as high as 1 mg/g fresh weight in the deficient leaves, compared with about 20 μg/g fresh weight for normal

Arginine		Citrulline		Ornithine
NH_2		NH_2		
$C{=}NH$		$C{=}O$		
NH		NH		NH_2
$\longrightarrow$ $(CH_2)_3$	$\rightleftharpoons$	$(CH_2)_3$	$\rightleftharpoons$	$(CH_2)_3$
$CH{-}NH_2$		$CH{-}NH_2$		$CH{-}NH_2$
$COOH$		$COOH$		$COOH$
$\downarrow$				$\downarrow$
NH_2		NH_2		
$C{=}NH$		$C{=}O$		NH_2
NH	$\longrightarrow$	NH	$\longrightarrow$	$(CH_2)_3$
$(CH_2)_3$		$(CH_2)_3$		$CH_2{-}NH_2$
$CH_2{-}NH_2$		$CH_2{-}NH_2$		
Agmatine		*N*-Carbamyl putrescine		Putrescine

FIG. 1. Metabolic pathways leading to putrescine formation.

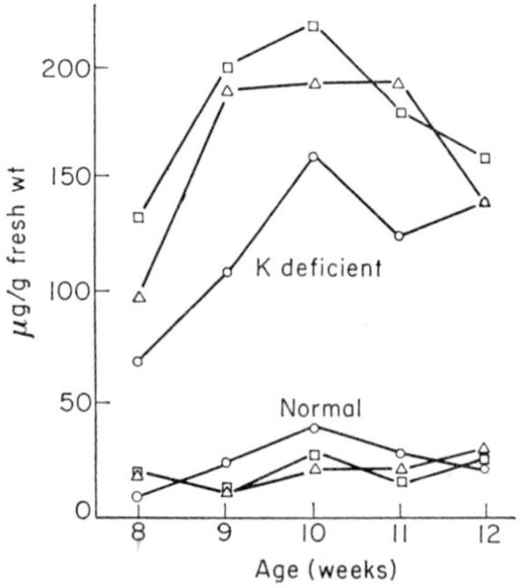

FIG. 2. Agmatine in three barley varieties.

barley leaves. The putrescine content of the deficient material was therefore five to ten times the amount of agmatine, but in the normal tissue it was at approximately the same level as the agmatine. The water content is increased by about 10% in the deficient plants and, in terms of the dry weight, the total nitrogen is reduced by 10% (Richards and Templeman, 1936). Therefore, in terms of total nitrogen there is an even greater difference in amine level with respect to the effect of treatments.

The accumulation of these amines under conditions of potassium deficiency appears to be a reaction of widespread occurrence in higher plants. A number of plant species were grown in normal and potassium-deficient conditions and the amine levels in the leaves were investigated (Table I). The groundsel did not show symptoms of potassium deficiency.

TABLE I

Effect of potassium deficiency on the amine level in leaves on a dry weight basis.

Increase in agmatine and putrescine	Increase in putrescine only	No increase detected
Tomato	Lettuce	Groundsel
Flax	Pea	
Cabbage	Beet	
Radish		
Red clover		
Oat		
Wheat		
Rye		

In order to confirm that agmatine is indeed the precursor of putrescine, solutions containing agmatine were fed to the cut shoots of barley seedlings for 24 hours and it was possible to show that putrescine is formed quite readily. Moreover, on feeding arginine, agmatine accumulated. Chromatography of the residual feeding solutions showed that bacterial activity was negligible.

In the ^{14}C arginine feeding experiments it was found that the rate of production of labelled putrescine was greater in potassium-deficient barley. Potassium therefore controls the speed of that part of the pathway lying between arginine and putrescine. Also, on chromatographing extracts of normal and potassium-deficient barley leaves it was found that both agmatine and putrescine were considerably increased in deficient plants, while the arginine content was scarcely affected. This indicated that the enzyme decarboxylating arginine to agmatine was probably increased by potassium deficiency.

In further investigations of this enzyme it was found that agmatine could be produced on incubating extracts of barley leaves with arginine in the presence of toluene as an antiseptic. It appears that this is the first record of the presence of arginine decarboxylase in higher plants. It was found that the activity of the enzyme could be conveniently determined by incubating dialyzed extracts with arginine and measuring the rate of agmatine formation. Barley plants were grown under normal and potassium deficient conditions and arginine decarboxylase activity was determined in extracts of the leaves. Table II shows the results for agmatine formation in these experiments with three varieties at two periods of growth. For the three varieties, the higher activity in each case is associated with the deficient leaves. Since the potassium-deficient plants have a reduced protein content, there would be an even greater divergence in enzyme activity, expressed on a protein basis, between the treatments.

TABLE II

The arginine decarboxylase activity in the leaves of three varieties of barley grown under normal and potassium-deficient conditions. Expressed as μg agmatine produced/h/g fresh weight.

Variety	Age (weeks)	Normal	Potassium-deficient
Proctor	6	62	172
	9	62	146
Mildew resistant	7	19	36
	10	25	72
Plumage Archer	8	78	101
	11	62	146

Considerable arginine decarboxylase activity was associated with the leaves of barley seedlings, none being detected in the roots. The enzyme was also apparently not located in the particulate fraction in the leaves. By treatment of crude extracts with acetone, or by freezing, much of the inactive protein could be removed, and the arginine decarboxylase was then recovered by precipitation with ammonium sulphate. Using this method the activity was increased six to ten times, with a recovery of up to 60 %. The carbon dioxide evolved on incubation of the preparation with arginine could then be measured manometrically. By this means it was possible to investigate some of the properties of the enzyme. It was found that out of seven L-amino acids, only arginine was attacked. The D-form of arginine apparently cannot be decarboxylated, and probably acts as a competitive inhibitor of the enzyme.

The properties of the barley enzyme may be compared with those of the arginine decarboxylase from *Escherichia coli* (Taylor and Gale, 1945).

The Michaelis constants are quite similar, but the pH optimum for the barley arginine decarboxylase is higher than for the bacterial enzyme. Also the barley enzyme is apparently more resistant to heat and inhibitors. Unlike the bacterial enzyme it was not found possible to separate pyridoxal phosphate as the coenzyme for the barley arginine decarboxylase, although it is expected that further work may show whether this is present.

It was of interest to determine whether putrescine was formed directly from agmatine with the production of urea, by an "agmatinase" which is analogous to arginase; or *via* *N*-carbamylputrescine, which is the decarboxylation analogue of citrulline (Fig. 1). In the experiments in which agmatine was fed to the cut shoots of barley seedlings, in addition to putrescine, *N*-carbamylputrescine was also detected on the chromatograms. This substance was shown to be easily converted into putrescine on being fed to barley seedlings (Smith and Garraway, 1964). It seems likely, therefore, that agmatine in barley is converted into putrescine with the intermediate formation of *N*-carbamylputrescine.

The enzyme converting agmatine to *N*-carbamylputrescine in barley could not be detected in barley leaf extracts, but it was possible to demonstrate this enzyme in extracts of other plants, for instance, sunflower cotyledons, or maize or cabbage leaves. It is possible that the barley enzyme is relatively unstable.

However, extracts of barley leaves were capable of converting *N*-carbamylputrescine to putrescine, the enzyme responsible for this being called *N*-carbamylputrescine amidohydrolase. The activity of this enzyme was measured by determining the ammonia produced when leaf extracts were incubated with and without added *N*-carbamylputrescine (Smith, 1965). Some ammonia was produced by the extract in the absence of added substrate and this was subtracted from the experimental estimations in order to obtain a true activity measurement (Fig. 3). This enzyme had an

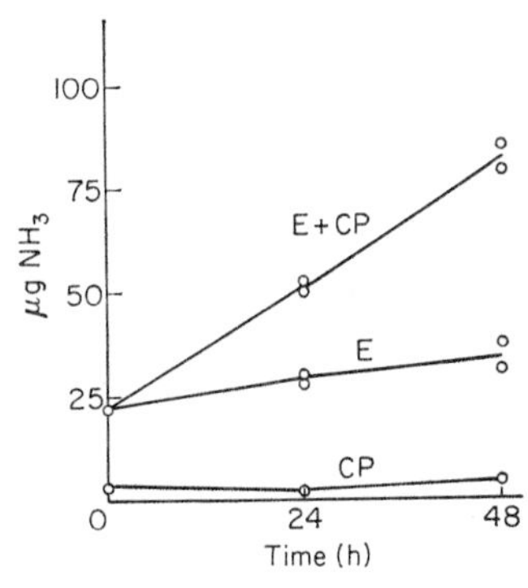

FIG. 3. Rate of ammonia formation by barley leaf extract (E) on incubation with *N*-carbamylputrescine (CP).

optimum pH of 7·5. It was sensitive to sulphydryl inhibitors, and the inhibition by *p*-chloromercuribenzoate was reversible by cysteine. The enzyme was found to be highly specific for *N*-carbamylputrescine and it would not attack dicarbamylputrescine, allantoin or citrulline. As it was known that phosphorolysis is involved in the enzymic degradation of citrulline (Krebs *et al.*, 1955), it was thought possible that phosphate might be required by the *N*-carbamylputrescine amidohydrolase. However, no reduction in activity could be found after prolonged dialysis against a phosphate-free buffer. Nor could any direct effect of the potassium ion be demonstrated.

An increased *N*-carbamylputrescine amidohydrolase activity was found in potassium-deficient plants (Table III), as was also found with the arginine

TABLE III

The *N*-carbamylputrescine amidohydrolase activity in the leaves of normal and potassium-deficient Proctor barley; expressed as μg NH_3 produced/h/g fresh weight.

Age of plant (weeks)	Normal	Potassium-deficient
11	18, 33, 27	89, 56, 64
14	53, 58	115, 108

decarboxylase activity. In this particular experiment however, the decarboxylase activity was increased more than 10-fold under conditions of potassium deficiency at the fourteenth week.

In order to explain the accumulation of amines in potassium-deficient plants, Coleman and Richards (1956) suggested that alkali-metal deficiency shifts the internal balance between inorganic anions and cations in the direction of increased acidity. It was therefore considered possible that the production of the basic amines functions as an internal compensating mechanism to maintain the pH at a constant value. In this context, the finding by Gale (1940) that the amino acid decarboxylases in *Escherichia coli* are more active when cultured in an acid medium is of particular interest. An attempt was therefore made to simulate the conditions of ionic imbalance experienced in potassium deficiency by feeding barley seedlings with hydrochloric acid through the roots (Smith and Sinclair, 1967).

Seedlings were grown for 7 days on butter-muslin supported by a layer of expanded polystyrene beads which floated in tap water. The water in one trough was then replaced by 0·025 N acid. Control seedlings were grown for 10 days in water. The arginine, agmatine and putrescine content of the leaves was increased with acid feeding, but other guanidino-compounds also appeared, and these made it difficult to estimate the relative levels of arginine and agmatine with accuracy. The shift in equilibrium resulting from an increased arginine level might be sufficient to account for an increase in putrescine, and this might occur even if the enzyme levels were to remain unchanged with acid feeding.

However, the arginine decarboxylase activity was found to be increased significantly with acid feeding on the bases of dry weight, total nitrogen and protein nitrogen (Fig. 4). The *N*-carbamylputrescine amidohydrolase

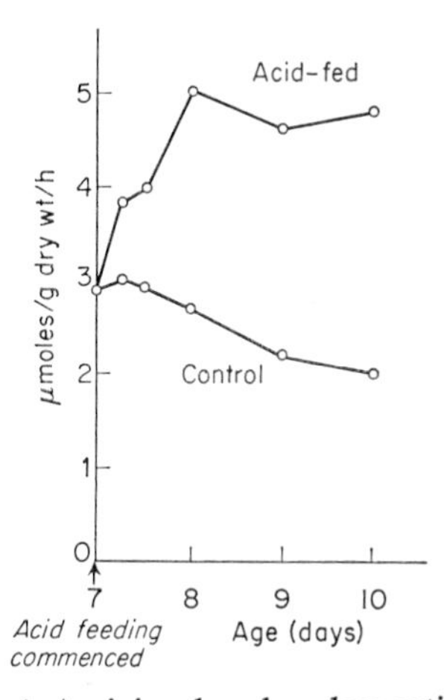

FIG. 4. Arginine decarboxylase activity.

activity was also increased with acid feeding, although this increase was not as great as that found with the decarboxylase (Table IV). No difference was found in the potassium content of the leaves and it seems possible that the effect was due to the increased hydrogen-ion concentration, since sulphuric acid feeding induced a similar rise in the activity of the arginine decarboxylase.

TABLE IV

The activities of enzymes concerned in putrescine formation in the leaves of barley seedlings grown ten days in water (control) or seven days in water, then three days in 0·025 N hydrochloric acid (acid-fed).

	Enzyme activity in μmoles/h/g protein		
	Control	Acid-fed	LSD at 5% level
Arginine decarboxylase	257	456	80
N-Carbamylputrescine amidohydrolase	146	207	49

It is probable therefore that a system involved in the synthesis of these enzymes is stimulated by a reduction of pH; moreover, in potassium deficiency there is a tendency towards increased acidity which accelerates the formation, or reduces the breakdown of these enzymes which, in turn, produce the basic amines and so restore the pH. However, the amines may be accumulated to levels at which they themselves become toxic, since Coleman and Richards (1956) found that a necrosis characteristic of potassium deficiency could be induced by feeding putrescine to normal barley leaves.

This system must be capable of achieving a very precise control, since in barley leaves, even in extreme potassium deficiency, it was not possible to detect a significant reduction in the pH of the cell sap.

SUMMARY

In barley leaves, putrescine is derived from arginine with the intermediate formation of agmatine and *N*-carbamylputrescine. In potassium-deficient barley there is an accumulation of agmatine and putrescine, and increased activities of arginine decarboxylase and *N*-carbamylputrescine amidohydrolase, the enzymes responsible for their formation. The activity of these enzymes is also increased on feeding inorganic acids to barley seedlings. It is considered possible that amine production is a response to internal conditions of increasing acidity and that this represents a new aspect of homeostasis in plants.

REFERENCES

Coleman, R. G. and Hegarty, M. P. (1957). *Nature, Lond.* **179**, 376.
Coleman, R. G. and Richards, F. J. (1956). *Ann. Bot.* **20**, 393.
Gale, E. F. (1940). *Biochem. J.* **34**, 853.
Krebs, H. A., Eggleston, L. V. and Knivett, V. A. (1955). *Biochem. J.* **59**, 185.
Linneweh, F. (1932). *Hoppe-Seyler's Z. physiol. Chem.* **205**, 126.
Miyaki, K. and Momiyama, H. (1956). *J. Biochem, Tokyo* **27**, 765.
Møller, V. (1955). *Acta. path. microbiol. scand.* **36**, 158.
Richards, F. J. and Coleman, R. G. (1952). *Nature, Lond.* **170**, 460.
Richards, F. J. and Templeman, W. G. (1936). *Ann. Bot.* **50**, 367.
Smith, T. A. (1963). *Phytochemistry* **2**, 241.
Smith, T. A. (1965). *Phytochemistry* **4**, 599.
Smith, T. A. and Garraway, J. L. (1964). *Phytochemistry* **3**, 23.
Smith, T. A. and Richards, F. J. (1962). *Biochem. J.* **84**, 292.
Smith, T. A. and Sinclair, C. (1967). *Ann. Bot.* **31**, 103.
Taylor, E. S. and Gale, E. F. (1945). *Biochem. J.* **39**, 52.

Section 2c

Some Properties of Plant Diamine Oxidase, a Copper-containing Enzyme

J. M. HILL AND P. J. G. MANN

Department of Biochemistry
Rothamsted Experimental Station, Harpenden, Herts., England

INTRODUCTION

The reactions catalyzed by the amine oxidases are generally represented by the equation:

$$RCH_2NH_2 + O_2 + H_2O \rightleftharpoons R.CHO + NH_3 + H_2O_2$$

which in the presence of catalase becomes

$$RCH_2NH_2 + \tfrac{1}{2}O_2 \rightleftharpoons R.CHO + NH_3$$

AMINE OXIDASES IN ANIMALS

The presence of monoamine oxidase and a diamine oxidase in animal tissues was established 30–40 years ago and these enzymes have been the subject of reviews by Zeller (1963) and Blaschko (1963). The substrate specificity of these two enzymes differs less than their names imply, but they differ considerably in their behaviour with inhibitors. Diamine oxidase, in contrast to monoamine oxidase, is inhibited by carbonyl reagents such as cyanide, semicarbazide and hydrazine, and this led to the suggestion that this enzyme contains a functional aldehydic or ketonic group (Zeller, 1938). The formation of hydrogen peroxide in the reactions catalyzed by the two amine oxidases suggested that both enzymes were flavoproteins but this has not been confirmed by later workers (Zeller, 1963).

Hirsch (1953) found a spermine oxidase in the plasma of ruminants and Blaschko and his co-workers described a related enzyme, benzylamine oxidase, in horse plasma (Bergeret *et al.*, 1957). Both these enzymes, like the diamine oxidase, are inhibited by carbonyl reagents.

AMINE OXIDASES IN PLANTS

The amine oxidases of higher plants have until recently received little attention. Cromwell (1943) suggested that *Atropa belladonna* contained an enzyme catalyzing the oxidation of putrescine. Werle and Zabel (1948) found a histaminase in plants from several families. Werle and Roewer (1950, 1952) reported the presence of a monoamine oxidase in a few other plants. Werle and Pechmann (1949) showed that this histaminase also catalyzed the oxidation of putrescine and cadaverine, and they claimed that it corresponded to the diamine oxidase of animal tissues. Like the animal enzyme, plant diamine oxidase was inhibited by carbonyl reagents and it was suggested that both pyridoxal phosphate and FAD were present. Goryachenkova (1956) gave further support to this suggestion but we have been unable to confirm it.

PEA-SEEDLING DIAMINE OXIDASE

Kenten and Mann (1952) described a diamine oxidase in pea seedlings that was not present in the dormant seed but developed rapidly after germination. Figure 1 shows that measurable diamine oxidase activity

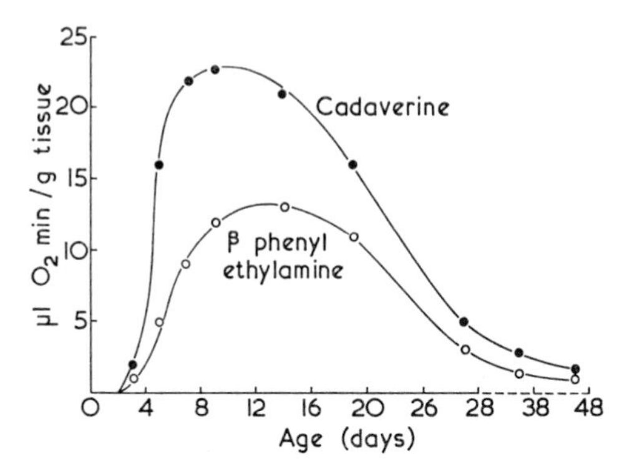

FIG. 1. Variation in rate of oxidation of cadaverine and β-phenyl-ethylamine with age of pea-seedling. (Kenten and Mann, *Biochem J.* **50**, 360)

first appeared 3–4 days after sowing, increased to a maximum over the period of 7–16 days and then decreased rapidly. The enzyme catalyzed the oxidation, not only of histamine and aliphatic diamines, but also of phenylalkylamines and, less rapidly, aliphatic monoamines and L-lysine.

Pea-seedling diamine oxidase catalyzes the oxidation of the diamino acids, ornithine and lysine, much less readily than diamines of the same hydrocarbon chain length. Clarke and Mann (1957a) showed that when the carboxyl groups were masked, as in the methyl esters, the rate of oxidation was greatly increased. Later, lysine peptides and other peptides were also shown to be oxidized by this enzyme. Clarke and Mann (1958) reported that the methyl esters of di- and tri-glycylglycine were substrates for diamine oxidase and suggested that a new group of plant amine oxidase substrates may be found among peptides.

TABLE I

Distribution of diamine oxidase activity in pea plants.

Part of Plant	Age of plant (days)			
	8	16	21	56
	Relative activity ($\mu l\ O_2$/min/g tissue)			
Whole plant	—	—	4·6	0·6
Stem (entire)	8·5	—	—	—
Shoot tip	—	—	12·0	30·0
Leaves	—	5	1·5	0
Stem (less shoot tip and/or leaves)	—	2·75	2·1	0
Seeds	—	—	—	0·2
Pods	—	—	—	0·2
Cotyledons	56	74·3	22·2	—
Roots (entire)	3	0·7	—	—
Roots (less nodules)	—	—	0·7	0·6
Nodules	—	—	22·0	39·0

Table I, based on the results of Kenten and Mann (1952) and later work, shows the distribution of the enzyme within plants of different ages. Though the total diamine oxidase activity in whole plants decreases with age, the enzyme is still very active in the shoot tips and root nodules of mature plants, suggesting that the enzyme may be localized in active meristematic regions.

Mann (1955) showed that diamine oxidase, which had been purified 300-fold, was inhibited by chelating agents and suggested that the enzyme might be a metalloprotein. Spectrographic analysis showed the presence of copper and manganese and a smaller amount of zinc and iron. The enzyme was purified further by column chromatography and Table II shows that, while the manganese diminished to trace amounts, the copper content increased to 0·08–0·09% (Mann, 1961).

Other work (Mann, 1961; Hill and Mann, 1962, 1964) showed that copper is essential for the activity of pea-seedling diamine oxidase. Treatment of the purified enzyme with sodium diethyldithiocarbamate removed the copper and, after dialysis to remove excess reagent, most of the original activity could be restored specifically by adding 1 μM Cu^{2+} ions (Fig. 2).

TABLE II

Copper and manganese content of the initial and final preparations of pea-seedling diamine oxidase (Mann, 1961).

Specific activity is defined as the amount (μmole) of putrescine oxidized/min/mg protein at 25°C.

Initial preparation			Final preparation		
Specific activity	Cu %	Mn %	Specific activity	Cu %	Mn %
16	0·05	0·07	49	0·08	<0·01
11	0·03	0·07	41	0·08	<0·01
14	0·04	0·06	41	0·09	<0·01

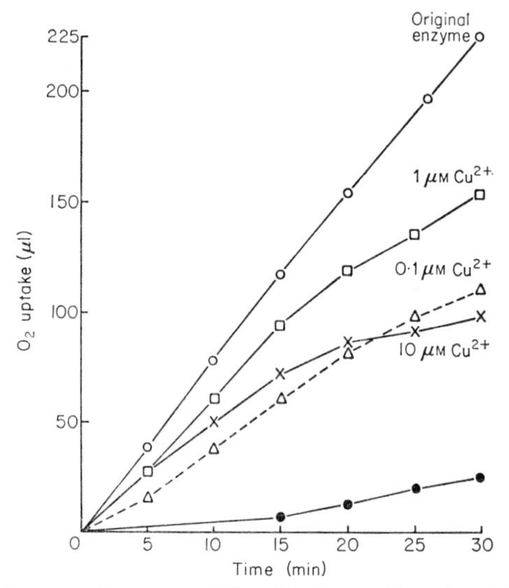

FIG. 2. Activation of copper-free pea-seedling diamine oxidase by Cu^{2+} ions; ●, Cu-free enzyme. (Mann, *Biochem. J.* **79**, 623)

When larger concentrations of Cu^{2+} ions were added, an inhibition was observed after the initial activation. Like ascorbate oxidase—another copper-containing enzyme—pea-seedling diamine oxidase is inactivated by free Cu^{2+} ions during the reaction it catalyzes. Figure 3 shows the inactivation of the diamine oxidase by 10- and 100 μM $CuSO_4$ during the oxidation of putrescine. Only Co^{2+} ions, of all the other metals tested, inactivated the enzyme and much greater concentrations were required of this than of copper. Hill and Mann (1962) also showed that copper in the pea-seedling diamine oxidase is in the divalent state and they obtained evidence that the reaction mechanism does not involve a copper-valency

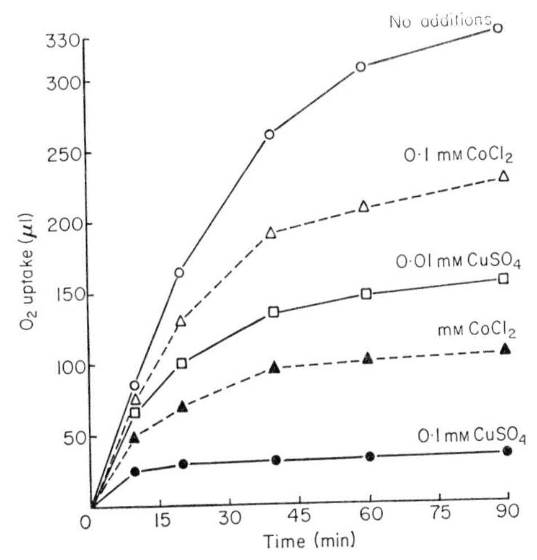

FIG. 3. Inactivation of pea-seedling diamine oxidase by Cu^{2+} and Co^{2+} ions.

change. Later, electron spin resonance showed that the copper present in plasma spermine oxidase is also divalent and that no valency change occurs during the reaction it catalyzes (Yamada *et al.*, 1963).

Purified pea-seedling diamine oxidase is pink and absorbs maximally at 500 nm (Fig. 4). The plasma amine oxidases (spermine oxidase and benzylamine oxidase), and an amine oxidase from *Aspergillus niger*, have been obtained in crystalline form as pink copper-containing proteins (Yamada and Yasunobu, 1962; Blaschko and Buffoni, 1965; Yamada *et al.*, 1965); a monomethylamine oxidase from *Trichosporon* is also pink and contains copper (Yamada *et al.*, 1966). The carbonyl compound in the prosthetic groups of the plasma amine oxidases has been identified as pyridoxal phosphate. The diamine oxidase of other animal tissues has been purified to give preparations (containing 0·06% copper) whose activities were increased by adding pyridoxal phosphate. It was therefore suggested that this enzyme's prosthetic group also contained pyridoxal phosphate and copper (Mondovi *et al.*, 1964). Blaschko and Buffoni (1965) claim that all the amine oxidases, including the plant diamine oxidase, which are inhibited by carbonyl reagents and contain copper, have pyridoxal phosphate in their prosthetic group. We have failed to show the presence of pyridoxal phosphate in hydrolysates of pea-seedling diamine oxidase and this, together with some differences between the properties of this enzyme and the other amine oxidases, suggests that its carbonyl component may not be pyridoxal phosphate.

Figure 4 shows that when substrate is added to pea-seedling diamine oxidase under anaerobic conditions, the colour changes from pink to yellow

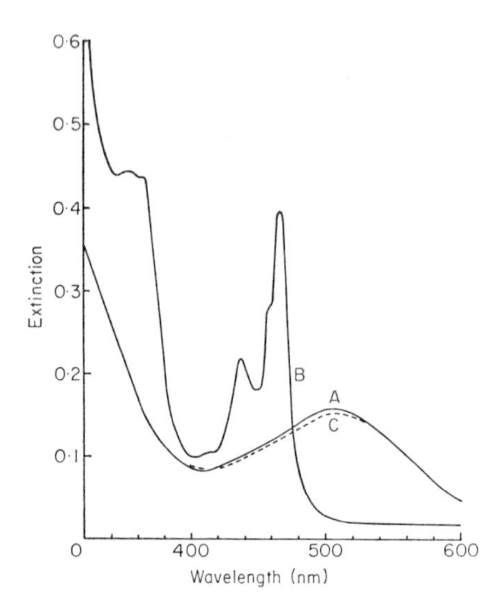

FIG. 4. Effect of putrescine on pea-seedling diamine oxidase under anaerobic conditions (A) Original oxidase, (B) A+0·1 ml 0·01M putrescine, (C) B after oxygenation for 2 min. (Mann, *Biochem. J.* **79**, 623; Hill and Mann, *Biochem. J.* **91**, 171)

and the absorption band at 500 nm is replaced by bands with maxima at 466, 437·5 and 350 nm; a slight inflexion also appears at 410–420 nm. On oxygenation the yellow colour changes to pink and the band at 500 nm reappears at its original intensity. We have attributed the yellow colour to the formation of an enzyme substrate complex that is unstable in air (Mann, 1961; Hill and Mann, 1964). Adding substrate to plasma benzylamine oxidase under anaerobic conditions reversibly discharges the pink colour but no yellow intermediate is formed (Buffoni and Blaschko, 1964). Similar results have been obtained with plasma spermine oxidase (Yamada and Yasunobu, 1962) and the amine oxidases obtained from *Aspergillus niger* (Yamada *et al.*, 1965) and *Trichosporon* (Yamada *et al.*, 1966). Yamada and Yasunobu (1962) attribute the bleaching of the plasma spermine oxidase by substrate under anaerobic conditions to the formation of a colourless enzyme substrate complex.

Figure 5 compares the absorption spectrum of a solution of the copper-free pea seedling diamine oxidase with that of the original enzyme. The copper-free protein is orange-pink and its colour is weaker than that of the original enzyme solution. When Cu^{2+} ions are added, the colour changes to pink and intensifies with the maximum absorption shifting from 480 to about 500 nm. The colour of the copper-free protein is discharged by substrate under anaerobic conditions but is not restored by oxygenation.

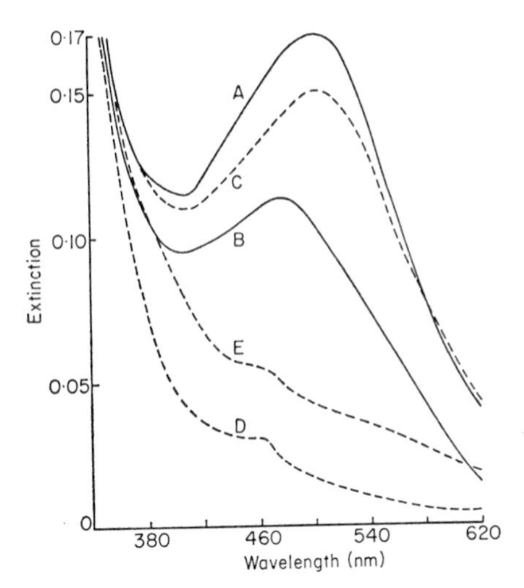

FIG. 5. Effect of Cu^{2+} ions and putrescine on the absorption spectrum of copper-free pea-seedling diamine oxidase; (A) diamine oxidase, (B) copper-free diamine oxidase, (C) B+1 μmole CuSO, after 30 min, (D) B+1 μmole putrescine under anaerobic conditions, (E) D after oxygenation for 2 min. (Hill and Mann, *Biochem. J.* **91**, 171)

The copper-free protein obtained from plasma spermine oxidase absorbs maximally at 380 nm. This band was attributed to the presence of pyridoxal phosphate and was unaffected by the addition of substrate under anaerobic conditions (Yamada and Yasunobu, 1962).

Table III compares the activity of pea seedling diamine oxidase with other purified amine oxidases. The plant enzyme is much more active than animal diamine oxidase and is also more active than the plasma amine oxidases towards benzylamine, one of the best substrates for these enzymes. These activities emphasize the differences between the plant diamine oxidase and amine oxidases obtained from other sources.

Hill and Mann (1964) found that the yellow enzyme-substrate complex formed between plant diamine oxidase and putrescine was stable under anaerobic conditions but became unstable when Cu^{2+} ions were added. Figure 6 shows the breakdown of the complex caused by 33 μM Cu^{2+} ions. The yellow complex had almost completely disappeared after 4 hours and the pink colour of the enzyme was only partially restored by oxygenation. Subsequent assay of the reaction mixture showed that most of the enzyme had been inactivated. We suggested that the Cu^{2+} ions cause an abnormal breakdown of the enzyme-substrate complex resulting in the inactivation of the enzyme. This may also explain the substrate-dependent inactivation of the enzyme by Cu^{2+} ions under aerobic conditions.

TABLE III

Comparison of the activities of some highly purified amine oxidases.

One unit of activity is defined as the amount of substrate oxidized (μmoles)/mg protein/min at the temperature stated. The published results have, in some cases, been recalculated to fit with this definition.

Enzyme	Substrate	Specific activity	Temp. °C	References
Pig kidney diamine oxidase	cadaverine	1·265	37	Mondovi *et al.*, 1967
	histamine	0·34	37	Mondovi *et al.*, 1964
Pea-seedling diamine oxidase	cadaverine	50·0	25	Hill and Mann, 1964
	histamine	2·7		
	benzylamine	2·2		
Histaminase	histamine	0·2	37	Kapeller-Adler and MacFarlane, 1963
Ox plasma spermine oxidase	benzylamine	0·12	37	Yamada and Yasunobu, 1962
Pig plasma benzyl-amine oxidase	benzylamine	0·18	37	Buffoni and Blaschko, 1964
Human plasma mono-amine oxidase	benzylamine	0·03	30	McEwen, 1965
Aspergillus niger amine oxidase	benzylamine	4·0	30	Yamada *et al.*, 1965

POSSIBLE PHYSIOLOGICAL ROLES OF DIAMINE OXIDASE

DIAMINE OXIDASE AND AUXIN FORMATION

Tryptophan has generally been considered as the precursor of indole-acetic acid in plants and two of the pathways proposed are shown in Fig. 7. The route involving indolepyruvic acid is most widely favoured but, as first shown by Skoog (1937), tryptamine can serve as an auxin precursor in some plants. Kenten and Mann (1952) showed that tryptamine is oxidized by pea seedling diamine oxidase and suggested that this enzyme may be involved in auxin formation. Clarke and Mann (1957b) identified indole-acetaldehyde as the product of this reaction. The recent work of Libbert *et al.* (1966) and Winter (1966) suggests that, under sterile conditions, plant tissue can convert tryptamine, but not tryptophan, to indoleacetic acid and they attribute the previous findings, that tryptophan can serve as a pre-cursor for auxin, to bacterial action.

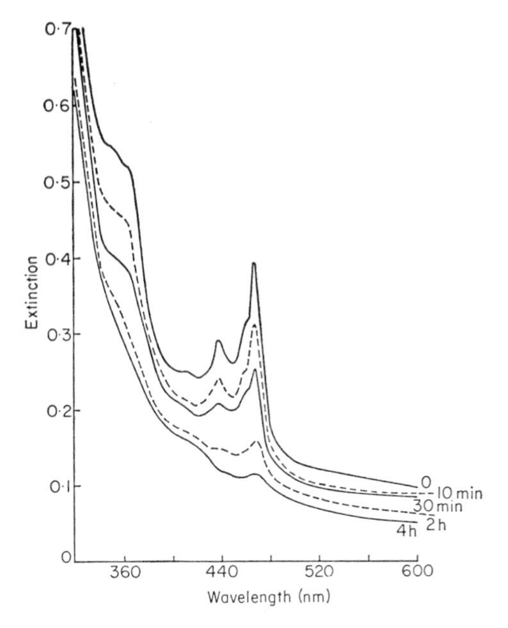

FIG. 6. Effect of Cu^{2+} ions on the pea-seedling diamine oxidase—putrescine complex under anaerobic conditions. Absorption spectra plotted at 0, 10 and 30 min and at 2 and 4 h after the addition of Cu^{2+} ions to the diamine oxidase—putrescine complex. (Hill and Mann, *Biochem. J.* **91**, 171)

Table I shows that diamine oxidase occurs in the active meristematic tissues of peas; as these are also the sites of auxin formation, this strengthens the possibility that amine oxidases may be involved in the formation of indoleacetic acid in plants. The amount of diamine oxidase necessary for auxin to be formed in most plants is not detectable by the normal manometric techniques. The large concentrations of this enzyme in peas and other legumes suggests that it may therefore have physiological roles other than that of auxin formation.

DIAMINE OXIDASE AND THE FORMATION OF HETEROCYCLIC COMPOUNDS

Tabor (1951) suggested that the amino aldehydes formed by the action of animal diamine oxidase on putrescine and cadaverine spontaneously cyclized to Δ^1-pyrroline and Δ^1-piperideine respectively; this was established by Mann and Smithies (1955a) and Hasse and Maisack (1955) in experiments with pea-seedling diamine oxidase. Mann and Smithies (1955a) showed the products of the oxidation, putrescine, cadaverine and lysine, were readily hydrogenated, forming derivatives of pyrrolidine, piperidine

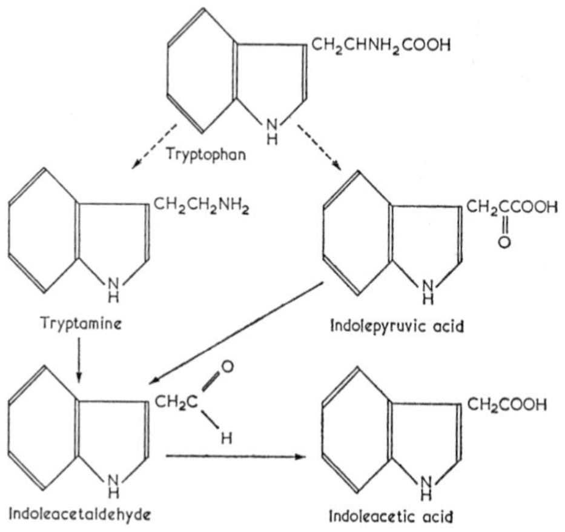

FIG. 7. Possible routes to the formation of indoleacetaldehyde from tryptophan.

and DL-pipecolinic acid, respectively. Figure 8 shows the reactions leading to the formation of pipecolinic acid from lysine; lysine is oxidized at the ϵ-amino group to give an aldimine which either cyclizes with the elimination of ammonia or reacts with water to give the aminoaldehyde which then cyclizes to Δ^1-piperideine-6-carboxylic acid. Macholan *et al.* (1967) reported that the product of oxidation of 2-hydroxyputrescine by animal and plant diamine oxidases, spontaneously forms pyrrole. Mann and Smithies (1955b) showed that the oxidation of *o*-aminophenylalkylamines may give cyclic products; 2-(2-aminophenyl)ethylamine was converted quantitatively to indole, presumably through the intermediate 2-(2-aminophenyl) acetaldehyde.

The formation of heterocyclic compounds as a result of the reactions catalyzed by diamine oxidase raises the question whether the enzyme is connected with the synthesis of alkaloids. Schöpf *et al.* (1953) showed that under certain conditions Δ^1-piperideine spontaneously forms tetrahydroanabasine. Hasse and Berg (1957, 1959) and Möthes *et al.* (1959) showed that pea seedling extracts can form anabasine from cadaverine, presumably by reduction of tetrahydroanabasine. Anabasine, an alkaloid occurring in *Anabasis aphylla* and tobacco, is an α-(β-pyridyl) piperidine that differs in constitution from nicotine by the replacement of the *N*-methylpyrroline ring by a piperidine moiety. Möthes *et al.* (1959) stated that this was the first enzymic synthesis of an alkaloid *in vitro*. Clarke and Mann (1959) showed that the cyclic products of the oxidase reaction with diamines

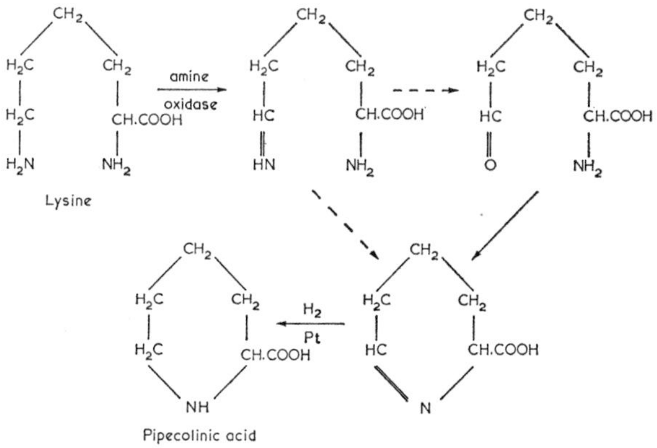

Lysine

amine oxidase

Pipecolinic acid

H_2 Pt

FIG. 8. Formation of pipecolinic (*syn.* pipecolic) acid from lysine.

combine spontaneously with β-oxoacids eliminating carbon dioxide; the alkaloids norhygrine and isopelletierine were formed when acetoacetate reacted with Δ^1-pyrroline and Δ^1-piperideine respectively.

Recent experiments with intact plants using radioactively labelled compounds show that amines can serve as precursors of many alkaloids. From some of these experiments, for example the formation of lupin alkaloids (Schütte *et al.*, 1964), it seems probable that the syntheses depend on the presence of diamine oxidase but it has not yet been proved that amines are the natural precursors of these alkaloids.

Although a large number of amines have been found in plants, they are usually present only in trace amounts. The findings by Richards and Coleman (1952) that putrescine accumulates in large amounts in potassium-deficient barley suggests that the role of amines in plant metabolism is greater than that expected from the amounts usually present. As diamine oxidase contains copper, it would be expected that copper-deficiency would make amines accumulate in species containing this enzyme. If tryptamine were to accumulate under these conditions, this would support the suggestion that diamine oxidase is involved in the formation of auxin from this compound.

REFERENCES

Bergeret, B., Blaschko, H. and Hawes, R. (1957). *Nature, Lond.* **180**, 1127.

Blaschko, H. (1963). *In* "The Enzymes" (P. D. Boyer, H. Lardy and K. Myrbäck, eds.) 2nd edn. Vol. 8, p. 337, Academic Press, London and New York.

Blaschko, H. and Buffoni, F. (1965). *Proc. R. Soc. B 163*, 45.
Buffoni, F. and Blaschko, H. (1964). *Proc. R. Soc. B 161*, 153.
Clarke, A. J. and Mann, P. J. G. (1957a). *Rep. Rothamsted exp. Stn for* 1956, 89.
Clarke, A. J. and Mann, P. J. G. (1957b). *Biochem. J.* **65**, 763.
Clarke, A. J. and Mann, P. J. G. (1958). *Rep. Rothamsted exp. Stn for* 1957, 100.
Clarke, A. J. and Mann, P. J. G. (1959). *Biochem. J.* **71**, 596.
Cromwell, B. T. (1943). *Biochem. J.* **37**, 722.
Goryachenkova, E. V. (1956). *Biokhimiya* **21**, 247.
Hasse, K. and Berg, P. (1957). *Naturwissenschaften* **44**, 584.
Hasse, K. and Berg, P. (1959). *Biochem. Z.* 331, 349.
Hasse, K. and Maisack, H. (1955). *Biochem. Z.* **327**, 296.
Hill, J. M. and Mann, P. J. G. (1962). *Biochem. J.* **85**, 198.
Hill, J. M. and Mann, P. J. G. (1964). *Biochem. J.* **91**, 171.
Hirsch, J. G. (1953). *J. exp. Med.* **97**, 345.
Kapeller-Adler, R. and MacFarlane, H. (1963). *Biochim. biophys. Acta* **67**, 542.
Kenten, R. H. and Mann, P. J. G. (1952). *Biochem. J.* **50**, 360.
Libbert, F., Wichner, S., Schiewer, U., Risch, H. and Kaiser, W. (1966). *Planta* **68**, 327.
McEwen, C. M. (1965). *J. biol. Chem.* **240**, 2003.
Macholan, L., Rozprimova, L. and Sedlackova, E. (1967). *Biochim. biophys. Acta* **136**, 258.
Mann, P. J. G. (1961). *Biochem. J.* **79**, 623.
Mann, P. J. G. and Smithies, W. R. (1955a). *Biochem. J.* **61**, 89.
Mann, P. J. G. and Smithies, W. R. (1955b). *Biochem. J.* **61**, 101.
Mondovi, B., Rotilio, G., Costa, M. T., Finazzi-Agro, A., Chiancone, E., Hansen, R. E. and Beinert, H. (1967). *J. biol. Chem.* **242**, 1160.
Mondovi, B., Rotilio, G., Finazzi-Agro, A. and Scioscia-Santoro, A. (1964). *Biochem. J.* **91**, 408.
Möthes, K., Schütte, H. R., Simon, H. and Weygand, F. (1959). *Z. Naturf.* **14B**, 49.
Richards, F. J. and Coleman, R. G. (1952). *Nature, Lond.* **170**, 460.
Schöpf, C., Braun, F. and Otte, K. (1953). *Ber. dt. chem. Ges.* **86**, 918.
Schütte, H. R., Sandke, G. and Lehfeldt, J. (1964). *Arch. Pharm. Berlin* **297**, 118; *Chem. Abst.* **60**, 13585g.
Skoog, F. (1937). *J. gen. Physiol.* **20**, 311.
Tabor, H. (1951). *J. biol. chem.* **188**, 125.
Werle, E. and Pechmann, E. V. (1949). *Justus Liebigs Annln Chem.* **562**, 44.
Werle, E. and Roewer, F. W. (1950). *Biochem. Z.* **320**, 298.
Werle, E. and Roewer, F. W. (1952). *Biochem. Z.* **322**, 320.
Werle, E. and Zabel, A. (1948). *Biochem. Z.* **318**, 554.
Winter, A. (1966). *Planta* **71**, 229.
Yamada, H., Adachi, O. and Ogata, K. (1965). *Agric. biol. Chem.* **29**, 912.
Yamada, H., Kumagai, H., Uwajima, T. and Ogata, K. (1966). *Mem. Res. Inst. Fd. Sci. Kyoto Univ.* **27**, 1.
Yamada, H. and Yasunobu, K. T. (1962). *J. biol. Chem.* **237**, 3077.
Yamada, H., Yasunobu, K. T., Yamano, T. and Mason, H. S. (1963). *Nature, Lond.* **198**, 1902.
Zeller, E. A. (1938). *Helv. chim. Acta* **21**, 880.

Zeller, E. A. (1963). *In* "The Enzymes" (P. D. Boyer, H. Lardy and K. Myrbäck, eds.) 2nd edn. Vol. 8, p. 313 Academic Press, London and New York.

Section 2d

Some Observations on the Specificity of Amino Acid Biosynthesis and Incorporation into Plant Proteins

L. FOWDEN, I. K. SMITH AND P. M. DUNNILL

Department of Botany, University College, London, England

INTRODUCTION

Our present information suggests that, as with other fundamental metabolic processes, the mechanism of protein synthesis in higher plants is basically similar to that already established in greater detail for several microbial and animal protein-synthesizing systems. The early activation reactions of the biosynthetic pathways are concerned with specifying the types of amino acids that may be incorporated into proteins, while the later genetically-directed steps determine the sequence in which amino acids become joined together. During the initial activation, amino acids are first converted into amino acyl-AMP-enzyme complexes in ATP-dependent processes catalyzed by a series of amino acyl-tRNA synthetase enzymes; the amino acyl residues are transferred subsequently to characteristic transfer RNAs (tRNAs) forming amino acyl-tRNA molecules. Once an amino acid residue has been attached to such a specific tRNA its subsequent incorporation into protein is normally ensured. Here, we shall consider three factors governing the selective incorporation of amino acids into plant protein, namely (a) the type of amino acid available to the protein-synthesizing system of plants and the specificity of biosynthetic processes elaborating them, (b) the ability of amino acyl-tRNA synthetases to combine with amino acids either as substrates or inhibitors, and (c) the selectivity of amino acyl group transfer to tRNAs as a function of the latter molecules.

THE AMINO ACID PRECURSORS

There is considerable evidence suggesting that the *free amino acid pool* of higher plant cells consists of several distinct metabolic pools (see Steward and Durzan, 1965). Certainly, individual free amino acids are not present in such cells in concentrations even remotely related to the amounts required for the elaboration of protein molecules, and situations exist in which carbon from exogenously-supplied ^{14}C-glucose enters protein molecules more rapidly than ^{14}C-atoms supplied in the form of amino acids (Steward *et al.*, 1958a). Such experiments, indicating the direct entry of glucose molecules into intermediary metabolic processes producing the amino acids utilized for protein synthesis, by inference also suggest that exogenously-supplied amino acids mainly enter sub-cellular pools from which substrates necessary for the amino acyl-tRNA synthetases are withdrawn only slowly. Presumably, the vacuole represents the site wherein a large proportion of these metabolically-discrete amino acid molecules are stored.

The possibility of such compartmentalization of amino acid metabolites has an important bearing on another aspect of protein-synthesizing systems. In addition to the twenty amino acid constituents of protein molecules nearly 150 other amino acids have been characterized as plant products. Of course any one plant species normally contains only a few of these additional substances, but many instances exist where particular substances are accumulated by plant tissues as the major free amino acid. The structures of certain of these additional amino acids are sufficiently like those of particular protein amino acids, that they may act as either substrates or inhibitors of the appropriate amino acyl-tRNA synthetase when *in vitro* enzyme studies are made (see below). Although it is not yet possible to cite an established example, one mechanism whereby a cell may prevent interference by *such amino acid analogues* with protein synthesizing systems could involve their strict sub-cellular separation from the activating enzymes.

Among the natural amino acid analogues, some represent homologues of protein amino acids, e.g. azetidine-2-carboxylic acid, I (of proline), and ethionine, II (of methionine). Other structural modifications yielding analogues include the replacement of atoms (or groups of atoms) within protein amino acids by others of similar size, e.g. canavanine, III (—O— replacing —CH_2— in arginine), or the presence of additional substituent groups, e.g. *m*-tyrosine, IV (a 3-hydroxy substituent on phenylalanine). Occasionally, effective analogues show more marked structural differences from the parent amino acid e.g. mimosine, V, a possible analogue of tyrosine and/or phenylalanine. The occurrence of these and other natural amino acid analogues has been recently reviewed (Fowden *et al.*, 1967).

The structural similarities discussed above leads one to speculate whether the biogenetic pathways giving rise to the natural analogues are modelled upon those producing the corresponding protein amino acids, even to the point of using some of the same enzymes. This possibility has been confirmed in part for the related biosyntheses of proline and the homologous pipecolic acid, which are both formed in rat liver, *Neurospora* and legume seedling extracts (Meister *et al.*, 1957; Yura and Vogel, 1959) by an identical set of enzymes. However, pipecolic acid is considered to be a constituent of only a small proportion of species forming the plant kingdom; if this view is correct, then it seems probable that the substrate specificities of the proline biosynthetic enzymes vary between plants and that one or more of these enzymes cannot utilize C_6 compounds. This last concept may operate more widely in the synthesis of homologous compounds, and has a firmly established parallel in the variable substrate specificity observed for the prolyl-tRNA synthetase from different plant species (Peterson and Fowden, 1965).

Although particular substances are characteristically produced by only a restricted range of plants, the control of biosynthesis could effectively rest upon the operation of only a few *specific* enzymes in the pathway. For instance, the formation of *m*-tyrosine by *Euphorbia* species may involve the normal early intermediates of the shikimate aromatic pathway, and differ only in the possession of an enzyme capable of either catalyzing an alternative type of transposition of the enolpyruvate residue of chorismic acid or hydroxylating phenylalanine at the *m*-position. Similarly, during mimosine biosynthesis, the precursor lysine (Hylin, 1964) may undergo some of the reactions leading to piperidine compounds (Ramstad and Agurell, 1964), but at a crucial point along the pathway an enzyme(s) specific to the producer-plants *Leucaena* and *Mimosa*, may divert a proportion of the original lysine residues into the unsaturated 3,4-dihydroxypyridine. The final step in mimosine biosynthesis may involve the condensation of dihydroxypyridine with serine; Tiwari *et al.* (1967) have indicated that serine provides the alanyl side-chain of mimosine. However, it is pertinent that serine has been postulated as a precursor of α,β-diaminopropionic acid (Reinbothe, 1962), and so the latter amino acid might be envisaged as providing both the alanyl side-chain and the ring-nitrogen atom of mimosine. If the above condensation step is involved, it may not require an enzyme specific to *Leucaena* and *Mimosa*, because the biosynthesis of a structurally related compound, β-pyrazol-l-ylalanine VI, from pyrazole and serine by cucurbits apparently utilizes a non-specific condensing enzyme that appears to be constitutive in all types of plant tested (Frisch *et al.*, 1967). Condensations of this type involving serine show a certain kinship with that involved in tryptophan formation from indole and serine.

The production of a large number of amino acids additional to the protein constituents therefore does not necessarily require the elaboration of a full set of biosynthetic enzymes by the plant for each unusual constituent; rather the operation of one (or a few) enzymes may channel metabolites into such compounds in particular plants. The coexistence of structurally related compounds, such as diaminopropionic acid, albizziine, mimosine and willardiine in species forming the Mimosoideae may owe their existence *in part* to a specific enzyme catalyzing diaminopropionate formation.

AMINO ACYL-tRNA SYNTHETASE SPECIFICITY

We shall examine the specificity of activation of protein amino acids and their natural analogues, dealing in detail with analogues of proline, arginine, and phenylalanine which have been examined in our laboratories.

The amino acids most widely studied in relation to the incorporation of proline into protein are azetidine-2-carboxylic acid, I, a lower homologue of proline and 4-hydroxyproline, VII. Azetidine-2-carboxylic acid constitutes up to 3% of the dry weight of the seeds of *Convallaria majalis L.* and occurs as an important nitrogenous constituent of many other Liliaceous plants (Fowden, 1956). This amino acid induces growth inhibition in *Escherichia coli* (Fowden and Richmond, 1963) and *Phaseolus aureus* (Fowden, 1963), but no growth inhibition occurs in the presence of proline. These growth inhibitions are accompanied by an apparent replacement of the proline residues in protein by azetidine-2-carboxylic acid (Fowden and Richmond, 1963). In contrast azetidine-2-carboxylic acid is absent from the protein of the azetidine-2-carboxylic acid producer species *Polygonatum multiflorum* which accumulates the amino acid to a concentration 50 times that which is lethal to *Phaseolus* (Peterson and Fowden, 1965). These authors examined the comparative specificity of the prolyl-tRNA synthetase from *Phaseolus* and *Polygonatum*. The prolyl-tRNA synthetase isolated from *Phaseolus* activated azetidine-2-carboxylic acid to 36% of the proline level, and the amino acid was transferred to a tRNA preparation at 3% of the proline level. The synthetase from *Polygonatum* exhibited greater specificity and only activated proline. The low efficiency of azetidine-2-carboxylic acid transfer suggests that toxicity may be attributable in part to the reduced rate of proline activation and incorporation as well as to the replacement of proline residues by analogues in protein molecules. It is the increased specificity of the prolyl-tRNA synthetase at the level of activation which protects *Polygonatum* from its own toxic product.

A particularly interesting proline analogue is 4-hydroxyproline which occurs in various structural proteins, e.g. collagen (Udenfriend, 1966) and the cell wall proteins of higher plants (Lamport and Northcote, 1960), but is not found as the free amino acid (*allo*-4-hydroxyproline does occur in some plants). The growth of phloem explants, which synthesize a protein rich in hydroxyproline, is inhibited by 4-hydroxyproline and the growth inhibition may be reversed by proline (Steward *et al.*, 1958b). Similarly, collagen contains a large amount of hydroxyproline which does not arise from free amino acid. The first step in collagen biosynthesis is thought to be the transfer of proline to tRNA by the prolyl-tRNA synthetase, and subsequent hydroxylation of either the prolyl-tRNA (Manner and Gould, 1963), or a ribosomal-bound proline-containing peptide or protocollagen (see review by Udenfriend, 1966). This example shows that the incorporation into protein of "additional" amino acids results from a modification of the products of the amino acyl-tRNA synthetases rather than from a modification of the specificity of the amino acyl-tRNAs themselves.

Canavanine is an amino acid analogue of arginine, formed by replacement of a —CH_2— group in arginine by an —O— atom. Canavanine constitutes up to 3% of the dry weight of the seeds of *Canavalia ensiformis* and has since been identified in various legumes (Bell, 1960; Birdsong *et al.*, 1960). Canavanine inhibits the growth of a wide range of organisms, (Richmond, 1959a; Steward *et al.*, 1958b; Graves *et al.*, 1962), and in all cases these growth inhibitions may be reversed by simultaneous addition of arginine. In addition to direct effects on arginine metabolism (Kalyankar *et al.*, 1958), canavanine has been reported to replace arginine residues in protein. For instance, when canavanine replaces arginine in the medium of the arginine-requiring mutant *Staphylococcus aureus* 524*SC* there is a synthesis of defective enzymes (Richmond, 1959a), resulting apparently from the replacement of arginine residues in the protein by canavanine (Richmond, 1959b). No canavanine is detected in the protein of cells grown in the presence of both arginine and canavanine. Canavanine inhibits the formation of arginyl-tRNA by the arginyl-tRNA synthetase isolated from *E. coli* (Boman *et al.*, 1961). This amino acid is also a substrate of the arginyl-tRNA synthetase from rat liver and causes a 50% inhibition of arginyl-tRNA formation. This marked inhibition is accompanied by the transfer of ^{14}C-canavanine to tRNA, and the incorporation of canavanine into protein apparently at the expense of arginine (Allende and Allende, 1964). The absence of canavanine from the protein of the canavanine producer species *Canavalia ensiformis* led to an investigation in our laboratories of the comparative specificity of the arginyl-tRNA synthetase from this producer species and a non-producer species (*S. aureus*). The indications from preliminary studies are that no significant activation of canavanine is brought about by the enzyme from either species. However, the levels of arginine activation observed to date are themselves low (J. Frankton and L. Fowden, unpublished results).

The amino acid mimosine, V, is markedly different in structure from any protein amino acid, but has been described variously as a tyrosine or phenylalanine analogue. Mimosine occurs in large quantities in the seeds of the legume species *Mimosa pudica* (Tiwari and Spenser, 1965), and *Leucaena leucocephala* (Hegarty and Court, 1964). This amino acid inhibits the growth of *Phaseolus aureus* and the inhibition is not reversed by co-addition of either tyrosine or phenylalanine (Smith and Fowden, 1966). This is in contrast to the observation in animals, where mimosine-induced growth inhibition may be partially reversed by either tyrosine (Crounse *et al.*, 1962) or phenylalanine (Lin *et al.*, 1964). As part of an investigation into the mechanism of mimosine toxicity in plants we examined the activation of mimosine by the tyrosyl- and phenylalanyl-tRNA synthetases from *Phaseolus*. These synthetases were purified by pH precipitation, ammonium sulphate fractionation, and DEAE-cellulose column chromatography. As shown in Fig. 1, the tyrosyl- and phenylalanyl-

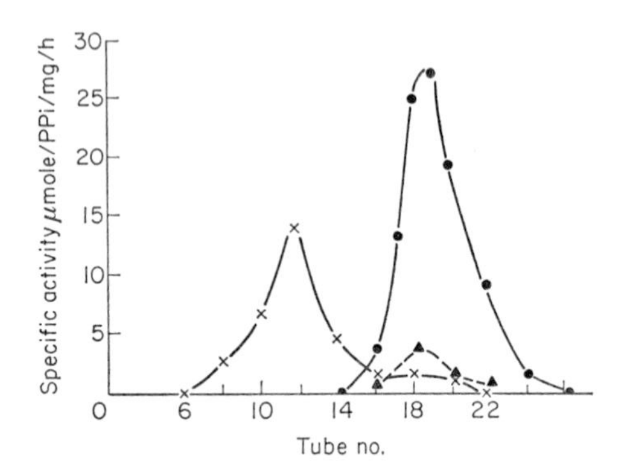

FIG. 1. Chromatography of tyrosyl- and phenylalanyl-tRNA synthetases on DEAE-cellulose. The protein was eluted from a 25 ml DEAE-cellulose column with a linear salt gradient (70 ml 0·1M Tris HCl versus 70 ml 0·1M Tris HCl-0·2M potassium chloride), 4 ml fractions were collected. ATP-PPi exchange in the presence of tyrosine ×; phenylalanine ●, and mimosine ▲.

tRNA activities are contained in different fractions of the DEAE-cellulose column eluate: mimosine is activated exclusively by the phenylalanyl-tRNA synthetase. Under comparable conditions (2·5 μmoles/ml amino acid substrates) mimosine-stimulated ATP-PPi exchange was 12% of the exchange with phenylalanine, which may be compared with 75% for *p*-fluorophenylalanine, a toxic synthetic analogue of phenylalanine; and 30% for *m*-tyrosine, a nontoxic natural phenylalanine analogue. The effect of mimosine, *m*-tyrosine and *p*-fluorophenylalanine on the transfer of ^{14}C-phenylalanine to tRNA by the phenylalanyl-tRNA synthetase from *Phaseolus* was also examined. The high level of ATP-PPi exchange supported by *p*-fluorophenylalanine and *m*-tyrosine was accompanied by a marked inhibition of phenylalanyl-tRNA formation. Mimosine on the other hand did not inhibit phenylalanyl-tRNA formation, and from the concentrations of mimosine and phenylalanine used it was inferred that mimosine was not transferred to tRNA. The structural differences between mimosine and phenylalanine therefore still permit mimosine activation by the phenylalanyl-tRNA synthetase but are sufficient to prevent any transfer of the amino acid.

Substitutions similar to those producing analogues of the larger amino acids already discussed would produce a more marked relative change in the overall size and shape of smaller amino acids. To date there is no evidence of analogues of the amino acids glycine, alanine, serine, threonine, cysteine, aspartic acid, and glutamic acid. The amino acid α-(methylenecyclopropyl)-glycine, VIII, occurs in the seeds of *Litchi* (Gray and Fowden, 1962). It markedly inhibits the growth of *Phaseolus* radicles and the inhibition may be reversed by co-addition of leucine (Fowden, 1963). This inhibition was not accompanied by any significant incorporation of the analogue into protein, but Peterson (1964) has shown that the amino acid is activated by the leucyl-tRNA synthetase isolated from *Phaseolus*. Ethionine, a higher homologue of methionine, has been identified in cellular extracts and in the growth medium of several bacteria (Fisher and Mallette, 1961). It produces growth inhibition in *E. coli* (Smith and Salmon, 1965), and *Coprinus lagopus* (Lewis, 1963), which may be partially reversed by co-addition of methionine. It apparently replaces methionine residues in the protein of various organisms (Levine and Tarver, 1951; Gross and Tarver, 1955; Lewis, 1963).

THE SELECTIVITY OF AMINO ACYL GROUP TRANSFER TO tRNA

The activation of non-protein amino acids and confirmation of their incorporation into protein in several laboratory experiments raises pointedly the question of why only twenty amino acids are universally employed in normal protein biosynthesis. It is clear that, while the activation step shows high specificity as between protein amino acids, a number of non-protein amino acid analogues can be activated to a significant degree. The only remaining step in which discrimination could occur is in the transfer of the activated amino acid to tRNA, since, as already mentioned, once bound to tRNA further incorporation into protein does not take note of the character of the amino acid. Whether the transfer to tRNA needs to exert an absolute specificity depends on (a) whether non-protein amino acids are available and accessible so that they must be dealt with and (b) whether it would make any difference to proteins if they were used instead of their protein amino acid equivalents. We will therefore first summarize the evidence on these two questions.

Availability could affect the choice of particular amino acids for protein biosynthesis in at least three ways; (a) the occurrence of the special 20 amino acids before others in pre-biological times could have disposed nature towards their use, (b) their biosynthesis might be more convenient, (c) the majority of amino acids might have been prevented from reaching the site of biosynthesis.

Experiments in recent years on the production of amino acids from simple mixtures such as methane and aqueous ammonia by thermal or electric discharge methods have shown that most protein amino acids can be so obtained (Harada and Fox, 1964). Although a specific search for the non-protein amino acids has not been made, the analyses indicate that some non-protein amino acids are inevitably produced at the same time. It is difficult to determine anything about early biosynthetic systems but the present day ones are evidently capable of making large quantities of non-protein amino acids. Azetidine-2-carboxylic acid, for example, can represent as much as 50% of the total nitrogen present in the rhizome of Solomon's Seal (Fowden, 1962). Non-protein amino acids such as ornithine, homoserine and homocysteine are common intermediates in the biosynthesis of protein amino acids so that unless the mechanism of biosynthesis has changed radically they must certainly always have been available. It was pointed out earlier that strict sub-cellular separation of non-protein amino acids may now occur. However, it is difficult to envisage an exclusion mechanism which functioned from the beginning so as to divide amino acids into two classes: rather it would appear to be a later development. In some plants the amino acyl-tRNA synthetase will not activate non-protein amino acid analogues, for example azetidine-2-carboxylic acid in *Polygonatum multiflorum* is not activated by the prolyl-tRNA synthetase enzyme. It seems reasonable that the latter represents a mutative change from a less specific enzyme since such absolute specificity is not a property of the prolyl-tRNA synthetases from plants which have no azetidine-2-carboxylic acid (Peterson and Fowden, 1965).

We have already noted several instances where protein and non-protein amino acids are very similar and when one looks in more detail at the structural character of naturally occurring amino acids it is not immediately obvious that the protein amino acids are special. Table I shows the protein amino acids and non-protein amino acids excluding particularly some of the more unusual ones. Side chains, representing R in $R{\cdot}CH(NH_2){\cdot}CO_2H$, of increasing size are plotted vertically and of different character horizontally, with protein amino acids underlined. Those non-protein amino acids which are substituted versions of protein amino acids are indicated by the extra substituents arrowed to the site of attachment. All protein amino acids are α-L-amino acids with the exception of proline, an α-L-imino acid and glycine, an optically inactive α-amino acid. They all have a hydrogen substituent on the α-carbon. There are a number of naturally occurring

TABLE I

Structural relationships between protein and some non-protein α-amino acid side-chains—R· in $R\cdot CH(NH_2)\cdot CO_2H$

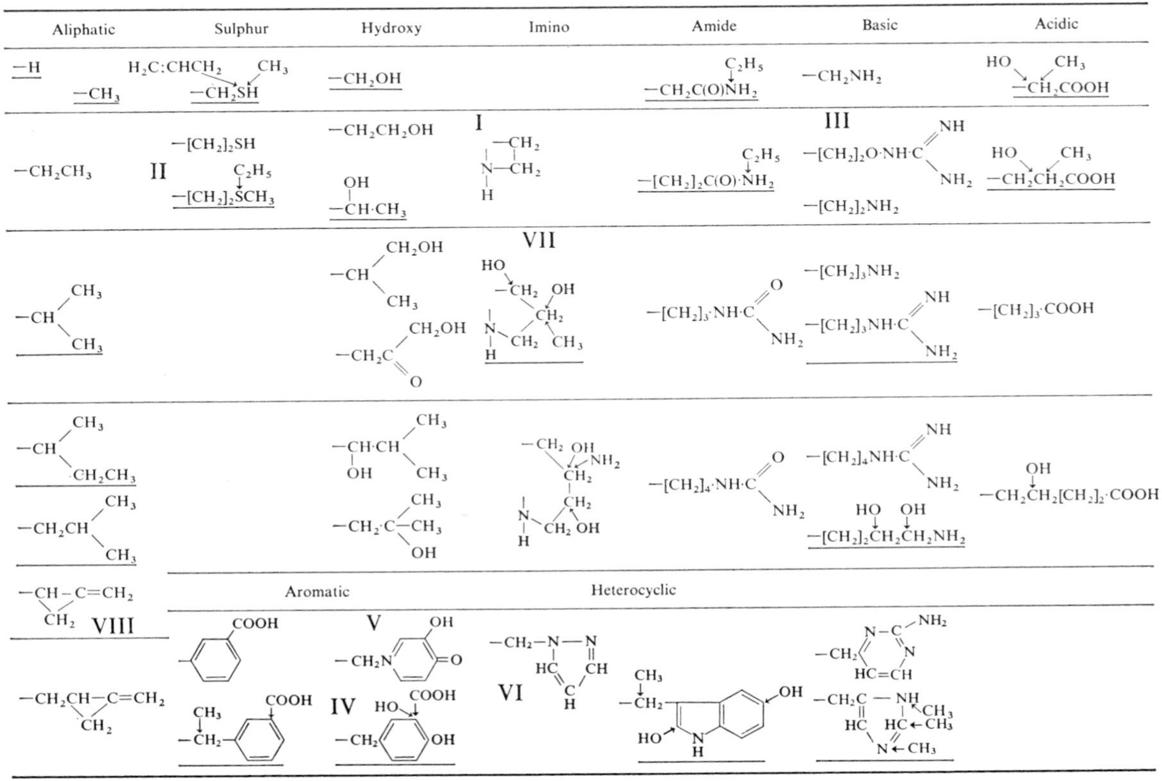

Aliphatic	Sulphur	Hydroxy	Imino	Amide	Basic	Acidic
$-H$; $-CH_3$	$H_2C{:}CHCH_2 \rightarrow -CH_2SH \leftarrow CH_3$	$-CH_2OH$		$-CH_2C(O)NH_2$ ($\leftarrow C_2H_5$ on N)	$-CH_2NH_2$	$-CH_2COOH$ ($\leftarrow HO$, $\leftarrow CH_3$)
$-CH_2CH_3$	$-[CH_2]_2SH$; II $-[CH_2]_2SCH_3$ ($\leftarrow C_2H_5$)	$-CH_2CH_2OH$; $-CH(OH)\cdot CH_3$	I $-CH_2$, $N-CH_2$, H (ring)	$-[CH_2]_2C(O)\cdot NH_2$ ($\leftarrow C_2H_5$ on N)	III $-[CH_2]_2O\cdot NH\cdot C(=NH)NH_2$; $-[CH_2]_2NH_2$	$-CH_2CH_2COOH$ ($\leftarrow HO$, $\leftarrow CH_3$)
$-CH(CH_3)CH_3$		$-CH(CH_2OH)CH_3$; $-CH_2C(=O)CH_2OH$	VII HO$\rightarrow$$-CH_2$, OH, CH_2, $N(H)$, CH_2, CH_3 (ring)	$-[CH_2]_3\cdot NH\cdot C(=O)NH_2$	$-[CH_2]_3NH_2$; $-[CH_2]_3NH\cdot C(=NH)NH_2$	$-[CH_2]_3\cdot COOH$
$-CH(CH_3)CH_2CH_3$; $-CH_2CH(CH_3)CH_3$		$-CH(OH)\cdot CH(CH_3)CH_3$; $-CH_2\cdot C(OH)(CH_3)-CH_3$	$-CH_2$, OH, NH_2, CH_2, CH_2, $N(H)$, CH_2, OH (ring)	$-[CH_2]_4\cdot NH\cdot C(=O)NH_2$	$-[CH_2]_4NH\cdot C(=NH)NH_2$; $-[CH_2]_2CH_2CH_2NH_2$ ($\leftarrow HO$, $\leftarrow OH$)	$-CH_2CH_2[CH_2]_2\cdot COOH$ ($\leftarrow OH$)
$-CH-C{=}CH_2$ (CH_2 ring) VIII	Aromatic		Heterocyclic			
$-CH_2CH-C{=}CH_2$ (CH_2 ring)	$-C_6H_4\cdot COOH$; $-CH_2(\leftarrow CH_3)-C_6H_4-COOH$	V $-CH_2N$ (pyridone ring, OH, =O); IV $-CH_2-C_6H_3$ (HO, COOH, OH)	VI $-CH_2-N$ (ring: N, HC, CH, C, H)	$-CH_2(\leftarrow CH_3)$ (indole ring, HO, OH, N–H)	$-CH_2$ (pyrimidine ring: $N-C-NH_2$, N, $HC{=}CH$); $-CH_2C$ (ring: NH$\leftarrow CH_3$, HC, $HC\leftarrow CH_3$, $N\leftarrow CH_3$)	

See text for explanation

non-α amino acids and others fully substituted at the α-carbon position. α-Amino acids are essential components of proteins as we know them and the question of what pseudo-proteins would be produced by the condensation of non-α-amino acids is rather too large for present discussion. Studies with molecular models indicate that substitution of the hydrogen at the α-carbon leads to considerable steric restriction of the normal side-chain and would interfere with the formation of important secondary structures such as the α-helix. It is evident that the protein amino acids are often the smallest members of the various functional classes; however, lysine and proline are not, and the aliphatic side-chain —CH_2CH_3 is not employed. There is a marked absence of all straight chain aliphatic amino acids in nature quite apart from proteins. Many of the non-protein amino acids are substituted forms of the protein ones and those of glutamic acid, proline and cysteine are particularly numerous. One might link this observation with the relative smallness of the majority of protein amino acids and ask whether there is an advantage in minimizing the size of the side-chain consistent with having a variety of chemical types. Most protein amino acids are monofunctional, i.e. hydroxyl or amino or carboxyl. One might think this to be the best structural choice for building a wide selection of enzymes because each functionally active unit could be selected, one at a time, for blending with others. Several non-protein amino acids possess polyfunctional side-chains, for example $[CH_2]_3 \cdot CH(NH_2)CO_2H$, and might be ruled out on this basis. However, they represent a small percentage of the total.

Since it is the protein molecules themselves that exhibit the unique biological properties, perhaps attention may be better directed to their structures rather than those of the constituent amino acids. Such structures must satisfy the requirements of stability in a fairly wide range of environments and possess catalytic activity in the case of enzymes. The structures of lysozyme (Blake *et al.*, 1965) and myoglobin, together with indirect evidence of other protein structures, indicate that the core of protein molecules must consist largely of non-polar residues and the outer shell of polar groups to confer stability. However the extensive studies of the tertiary structure of haemoglobins in relation to amino acid sequence, in a variety of living organisms (Perutz *et al.*, 1965), indicate that closely similar tertiary structures and essentially identical catalytic function can be produced in spite of the variation of all but a very few, and possibly of all, residues. The only reservation is that non-polar residues in the interior of the molecule must almost always be replaced by other non-polar ones, but even here the latitude is great. Taken with Kendrew's finding that the number of Van der Waal's contacts made by identical types of non-polar residues in different parts of the myoglobin molecule varies quite widely

(Kendrew, 1966), one may conclude that non-polar, non-protein amino acids could apparently serve the same function equally well. Too little is yet known to permit detailed discussion of the ways in which the catalytic activity of enzymes is dependent upon the character of polar residues and so it is impossible to know whether non-protein amino acids could serve equally well. One can point to near analogues which could presumably act in a similar manner; for example, ornithine for lysine, homoarginine for arginine, α-aminoadipic acid for glutamic acid, and homoserine for serine. However in myoglobin the polar groups of serine and aspartic acid appear occasionally to form hydrogen bonds with main chain peptide groups several residues away. Polar groups at the end of longer side-chains would be unable on steric grounds to form hydrogen bonds, at least with nearby peptide groups. Histidine has no near analogues. The presence of a proline residue is always associated with a break in the α-helix, accompanied by chain bending, and azetidine-2-carboxylic acid and pipecolic acid, though structurally very similar to proline, would act in markedly different ways in these respects. However the globin studies (Perutz *et al.*, 1965) indicate that the role of proline can be fulfilled by a variety of alternative arrangements of residues so that an irreplaceable role for proline cannot be established.

The above arguments suggest that proteins could be constructed at least partially from non-protein amino acids and the discussion of availability has indicated that, at least initially, protein and non-protein amino acids were available and accessible. Both these conclusions lead to the expectation that one step in incorporation, evidently the amino acyl group transfer to tRNA, must be highly specific: this appears to be the case.

The assignment of protein amino acids to the available codons of messenger RNA, i.e. the genetic code, is now known for a variety of organisms and appears to be universal. Transfer-RNA molecules binding a particular amino acid contain a trinucleotide sequence, the anticodon, which is complementary to the codon for this amino acid. Broadly-speaking there is therefore a complementary genetic code for anticodons and amino acids though unusual bases may be substituted in some instances. The central question then arises as to whether there is a direct stereochemical relation between amino acids and the tRNA-anticodon which defines the selection of twenty amino acids or whether no such relationship exists. Assuming the latter view, the use of twenty amino acids was arrived at by evolution either (a) after trying all amino acids or (b) by using a few initially and adding others. Alternative (a) is not readily reconciled with the above examination of amino acid structures which indicates no sharp division of character between those selected and others but evolution might select for very subtle differences. The inevitable production of non-protein amino acids when primitive chemical conditions are simulated throws doubt on alternative (b). Regarding steric interaction between amino acids and trinucleotides, the anticodon trinucleotide appears to represent the most likely selector for amino acids (Dunnill, 1966). Direct interaction of amino acids with the codons of messenger RNA or with the same codon sequence in tRNA (Pelc and Welton, 1966) now seems less likely as the mechanism of biosynthesis is better understood and further tRNA sequences become known (Crick, 1966; 1967). Model building studies can rarely be conclusive but such work in our laboratories supports stereochemical selection of amino acids by anticodon-trinucleotide cages. Studies are in progress to examine how this permits access of the relevant protein amino acids to the terminal-CCA end of tRNA for chemical linkage (P. M. Dunnill and A. C. R. Samson, unpublished). Steric selection is consistent with the known degeneracy of the code and with the concept of "wobbliness" put forward by Crick (1966). It must be reconciled with the role of the amino acyl-tRNA synthetases which certainly exhibit some specificity for amino acids and transfer RNA's. The studies of non-protein amino acid activation which have been mentioned indicate however that the first step exerts a screening effect but is not fully specific and there is as yet no way of knowing whether the specificity of the second step rests largely with the amino acid or the enzyme holding it.

CONCLUSION

To conclude, it is evident that the study of non-protein amino acids in plants is highlighting a number of problems associated with amino acid and protein biosynthesis: the specificity of enzymes synthesizing amino acids; the organization of amino acid pools; the screening and specific selection of amino acids for incorporation into protein. These problems are being studied intensively in many laboratories and in view of the universal nature of the genetic code and of many biosynthetic pathways their solution can be expected to have important implications in the study both of plants and of other organisms.

REFERENCES

Allende, C. C. and Allende, J. E. (1964). *J. biol. Chem.* **239**, 1102.

Bell, E. A. (1960). *Biochem. J.* **75**, 618.

Birdsong, B. A., Alston, R. and Turner, B. L. (1960). *Can. J. Bot.* **38**, 499.

Blake, C. C. F., Koenig, D. F., Mair, G. A., North, A. C. T., Phillips, D. C., and Sarma, V. R. (1965). *Nature, Lond.* **206**, 757.

Boman, H. G., Boman, I. A. and Maas, W. K. (1961). *In* "Biological Structure and Function" (T. W. Goodwin and O. Lindberg, eds.) Vol. 1, p. 297, Academic Press, New York.

Crick, F. H. C. (1966). *Nature, Lond.* **212**, 1397. (Report of a lecture, see corrections in Crick, 1967).

Crick, F. H. C. (1967). *Nature, Lond.* **213**, 119.
Crounse, R. G., Maxwell, J. D. and Blank, H. (1962). *Nature, Lond.* **194**, 694.
Dunnill, P. (1966). *Nature, Lond.* **210**, 1267.
Fisher, J. F. and Mallette, M. F. (1961). *J. gen. Physiol.* **45**, 1.
Fowden, L. (1956). *Biochem. J.* **64**, 323.
Fowden, L. (1962). *Endeavour* **21**, 35.
Fowden, L. (1963). *J. exp. Bot.* **14**, 387.
Fowden, L. and Richmond, M. H. (1963). *Biochim. biophys. Acta* **71**, 459.
Fowden, L., Lewis, D. and Tristram, H. (1967). *Adv. Enzymol.* **29**, 89.
Frisch, D. M., Dunnill, P. M., Smith, A. and Fowden, L. (1967). *Phytochemistry* **6**, 921.
Graves, D. J., Boldt, R. E. and Boyer, P. D. (1962). *Biochim. biophys. Acta* **57**, 165.
Gray, D. O. and Fowden, L. (1962). *Biochem. J.*, **82**, 385.
Gross, D. and Tarver, H. (1955). *J. biol. Chem.* **217**, 169.
Harada, K. and Fox, S. W. (1964). *Nature, Lond.* **201**, 335.
Hegarty, M. P. and Court, R. D. (1964). *Aust. J. agric. Res.* **15**, 165.
Hylin, J. W. (1964). *Phytochemistry* **3**, 161.
Kalyankar, G. D., Ikawa, M. and Snell, E. E. (1958). *J. biol. Chem.* **233**, 1175.
Kendrew, J. C. (1966). *In* "Principles of Biomolecular Organization" (G. E. W. Wolstenholme, ed.) a Ciba Symposium, p. 89.
Lamport, D. T. A. and Northcote, D. H. (1960). *Nature, Lond.* **188**, 665.
Levine, M. and Tarver, H. (1951). *J. biol. Chem.* **192**, 835.
Lewis, D. (1963). *Nature, Lond.* **200**, 151.
Lin, K. T., Lin, J. K. and Tung, T. C. (1964). *J. Formosan med. Ass.* **63**, 10.
Manner, G. and Gould, B. S. (1963). *Biochim. biophys. Acta* **72**, 243.
Meister, A., Radhakrishnan, A. N. and Buckley, S. D. (1957). *J. biol. Chem.* **229**, 789.
Pelc, S. R. and Welton, M. G. E. (1966). *Nature, Lond.* **209**, 868.
Perutz, M. F., Kendrew, J. C. and Watson, H. C. (1965). *J. molec. Biol.* **13**, 669.
Peterson, P. J. (1964). Ph.D. Thesis, University of London.
Peterson, P. J. and Fowden, L. (1965). *Biochem. J.* **97**, 112.
Ramstad, E. and Agurell, S. (1964). *A. Rev. Pl. Physiol.* **15**, 143.
Reinbothe, H. (1962). *Flora, Jena.* **152**, 545.
Richmond, M. H. (1959a). *Biochem. J.* **73**, 155.
Richmond, M. H. (1959b). *Biochem. J.* **73**, 261.
Smith, I. K. and Fowden, L. (1966). *J. exp. Bot.* **17**, 750.
Smith, R. C. and Salmon, W. D. (1965). *J. Bact.* **89**, 687.
Steward, F. C., Bidwell, R. G. S. and Yemm, E. W. (1958a). *J. exp. Bot.* **9**, 11, 285.
Steward, F. C., Pollard, J. K., Patchett, A. A. and Witkop, B. (1958b). *Biochim. biophys. Acta* **28**, 308.
Steward, F. C., and Durzan, D. J. (1965). *In* "Plant Physiology" (F. C. Steward, ed.) Vol. 4, p. 379, Academic Press, New York
Tiwari, H. P., Penrose, W. R. and Spencer, I. D. (1967), *Phytochemistry* **6**, 1245.
Tiwari, H. P., and Spenser, I. D. (1965). *Can. J. Bot.* **43**, 1687.
Udenfriend, S. (1966). *Science, N.Y.* **152**, 1335.
Yura, T., and Vogel, H. J. (1959). *J. biol. Chem.* **234**, 335, 339.

Discussion on Section 2

Following the paper by Professor Davies, Dr E. J. Hewitt referred to the problems of artefacts and allosteric enzymes and mentioned a recent short paper in *Biochem. J.* by Rabin who had drawn attention to the fact that this type of kinetics may be produced by conformational changes that are not associated with the oligomeric series at all. This was therefore an independent and very important alternative approach to this problem. He asked if there are many artefacts in macromolecular biochemistry itself. Although one of the earliest systems which was accepted in this context was ascorbic acid oxidase, where a number of proteins and some copper could have ascorbic acid oxidizing capacity, he wondered whether pyridoxyl phosphate proteins might behave as hydroxylamine reductases. His own HR_2 preparations which have a big peak at 399 nm might in fact contain pyridoxyl phosphate, but this could constitute an artefact producing an oxime of pyridoxyl which reacts only with BV·, and he queried whether this was a common situation.

Professor Davies replied: "Biochemists are anthropomorphic in their arguments and are lovers of teleology. The classical example is D-amino acid oxidase from animal tissue; D-amino acid oxidase is found in greater abundance than the L- form, but it has no apparent function. No biochemist wants to believe that his particular enzyme has only a test-tube function, and a function may be found at a later date. For example, the function of the widely distributed enzymes, glyoxylase I and II, has long been obscure, but Szent-Gyorgyi has recently proposed that they may have a role in cell division. On the other hand, many enzymes show broad specificity and will react readily with non-physiological substrates".

Professor F. R. Whatley asked about the significance of the fact that the only way ammonia enters the cycle is through glutamic dehydrogenase. Professor Davies suggested that it is the "economy of tools", doing a job of work with the minimum of effort.

Replying to a further query from the Chairman as to why he threw so much scorn on feedback mechanisms, Professor Davies said: "I am sorry if I created that impression. One has to beware of the band wagon. A curve that appears to be allosteric, i.e. a sigmoid curve, is by the simplest explanation a second order reaction. One should eliminate the simplest explanation first."

Recent work has indicated that a number of control enzymes are subject to end-product inhibition. Kinetic studies of these enzymes have shown that they may be classified as "allosteric K" or "allosteric V" systems according to whether they give a sigmoid curve in the plot V/S or V/I (Monod *et al.*, 1965).

There seems to be a tendency to assume that the demonstration of sigmoid kinetics implies that the enzyme concerned must be a control enzyme with allosteric properties. I wish to stress that there are alternative explanations of sigmoid kinetics and the possibility of an nth order reaction should be considered. Thus pyruvic decarboxylase exhibits a sigmoid relationship in the plot V/S and an explanation can be offered on the assumption that the reaction is second order.

Conversely there is a tendency to assume that control enzymes must show sigmoid kinetics. Mr Slaughter and I have examined 3-phosphoglycerate dehydrogenase from peas and found that this enzyme is inhibited by L-serine (end-product inhibition). When extracts have been "aged" the enzyme exhibits the kinetic properties of an "allosteric V" system. However, when the enzyme is freshly prepared, serine produces simple non-competitive inhibition. In this case it could be argued that the "allosteric V" kinetics represent an artefact of isolation".

Arising from Dr T. A. Smith's contribution, Dr Hewitt drew attention to the fact that experiments were initially carried out on the replacement of potassium in plants by sodium, rubidium and calcium; and asked if this had been done in the context of the author's recent studies. Dr Smith stated that this would be interesting, but had not been done. Dr Hewitt pointed out that the role of potassium included effects on the pyruvic kinase system and acetic thiokinase system in higher plants. He asked if there was any possibility that the products formed in the forward reactions directed by ATP in these systems would in any way affect the activity of the enzymes involved in these investigations. Dr Smith doubted if the stage has been reached when it could be said for sure how the products of these reactions might affect the activity of the enzymes involved in amine formation. In reply to the Chairman's query as to the effect of the hydrogen ion concentration of the external medium on the potassium availability, Dr Smith said that there was no considerable change in the internal potassium levels.

Professor Whatley also enquired about the effect of different nitrogen sources on the amount of putrescine formed, and was informed that when barley is grown in an ammonium medium there is more formed than when it is grown in a nitrate medium.

Dr G. G. Freeman congratulated the author on the elucidation of the mechanism of formation of putrescine in potassium-deficient plants and remarked that the effect of the element in controlling plant growth remained to be explained. He referred to the work of J. Dicks and D. Tempest on *Aerobacter aerogenes* and their conclusion that the relatively large amounts of potassium found in the organism were required to maintain ribosomal structures in a suitable conformation for protein synthesis.

Mr E. A. Kirkby commented: "One might anticipate that 50% of the cations of barley are present as potassium. If putrescine is important as a compensating cation in potassium deficient plants, considerable amounts would have to be accumulated unless anion concentrations were also reduced. Has a balance sheet been drawn up taking into account putrescine, the organic acids and inorganic anions and cations?"

Dr Smith expanded this by stating: "A reduction of putrescine content arises when rubidium or sodium is fed to potassium-deficient barley. Rubidium is more effective than sodium for this (Coleman and Richards, 1956). Regarding the second question, we still have very little information on the levels of the organic acids in potassium-deficient barley leaves and we cannot provide an answer for this at present."

The discussion on the contribution of J. W. Hill and P. G. Mann commenced with a reply to the Chairman in which Mr Hill stated that most experiments had been performed in orthophosphate buffers but both veronal and pyrophosphate buffers had also been used. In veronal buffers, lower concentrations of Cu^{2+} ions than in orthophosphate buffers inhibit the diamine oxidase; pyrophosphate buffer prevents Cu^{2+} ion inhibition of the enzyme, presumably by the formation of an inactive copper-pyrophosphate complex. Professor Whatley also enquired if diamine oxidase were universal in its distribution. Mr Hill thought this was possible, but not confirmed. To a supplementary point about auxins, he said his comments had been tentative and were not definitive about the way in which auxins are formed.

Dr Hewitt asked if the authors had observed any fine structure in the region of 255 to 285 nm in the absorption spectrum of diamine oxidase to which Mr Hill answered: "We have never seriously looked at the fine structure in this region as traces of other proteins could significantly affect the result. Our preparations of diamine oxidase have an absorption maximum at 280 nm with a slight inflexion at 290 nm." Mr Hill further confirmed Dr Hewitt's suggestion that if they had a highly purified protein sample it would be worth looking at.

Mr Hill, in reply to Dr Rosalie Cox said that no work had been done on levels of diamine oxidase in plants grown with potassium deficiencies.

After Mr I. K. Smith's presentation of "Some observations on the specificity of amino acid biosynthesis and incorporation into plant proteins" he expanded, at the Chairman's suggestion, on "compartmentalization" by saying: "In certain instances compartmentalization may be the method by which the producer-plant avoids the toxicity of its own product; however, in certain plants this would not appear to be the case. For instance, in canavanine-producer species, canavanine is distributed throughout the plant, its concentration in various organs is subject to diurnal and seasonal variations, and it constitutes a high percentage of the soluble nitrogen present in phloem explants; this suggests that it is an essential constituent of the amino acid pool which is utilized during growth. I think that in such canavanine-producer species there would be an increase in the specificity of the enzymes concerned with arginine metabolism so that canavanine was no longer a substrate. This is supported by the work of Peterson and Fowden mentioned earlier; they showed that the absence of azetidine-2-carboxylic acid from the protein of producer species was due to the increased specificity of the prolyl-tRNA synthetase as compared with the enzymes from sensitive species.

I should add that certain toxic amino acids are stored in a non-toxic form, for instance, the toxic effects of γ-glutamyl-β-aminopropionitrile (the *Lathyrus* factor) are only observed when the free amino acid is liberated by hydrolysis of the peptide."

Professor Whatley further commented: "You seemed to indicate that it was technically difficult to decide whether an amino acid analogue was incorporated into the end group of the protein or whether it was in the middle of a protein. Wouldn't it be fairly easy to treat with dinitrofluorobenzene to determine the N-terminal group and with bromosuccinamide for the C-terminal group." Mr Smith replied that he had not intended to give the impression that it was difficult. His statement was a criticism of the use of the presence of an analogue in the hydrolysate of trichloroacetic acid-precipitated protein as a criterion of incorporation, since this procedure does not exclude the possibility that the analogue terminates protein synthesis, and is present only at the ends of partially completed polypeptides. End-group analysis would readily establish whether this was true or not.

To a further query from the Chairman regarding anything known about the effects of these analogues on metabolism of animal tissue, Mr Smith said that azetidine-2-carboxylic acid inhibits the development of chick embryos. Several reports show that canavanine is inhibitory to animal growth (mice, rats, and rabbits) particularly with regard to the rate of body weight increase; and animals maintained on a diet consisting only of *Canavalia* (a canavanine-producer species) die prematurely.

Professor Whatley pointed out that in the preprint the authors had suggested that enzyme activity is largely related to tertiary structure and that the primary structure can be modified to some extent without affecting activity. Mr Smith, in explanation, said: "This is based on the work of Perutz and Kendrew on haemoglobin which suggests you can modify all but eight of the amino acids in the haemoglobin and still get precisely the same function. Although perhaps it is not fair to call haemoglobin an enzyme, its function is so close to one. I think it is possible to change all the amino acids in an enzyme apart from these eight".

In reply to the question whether proline was one of them, he said: "You can change all the proline in haemoglobin and replace it by other groups or residues which perform the same function of terminating units. The views on proline are not as strong as when the first structure was determined. You can change most of the residues, but not one in the series involved in the active site. You would not do any good to change that for some other. I think it is true to say that proline is the neatest way of terminating an α-helix but not the only way."

Dr Hewitt made the following contribution: "B. A. Notton and myself have found that L-azetidine-2-carboxylic acid is an inhibitor of nitrate reductase induction. This enzyme, as you know, can be induced in leaf fragments of plants grown in the absence of nitrate or molybdenum by infiltration of the appropriate substrate or metal. We have found that the inhibition of this induction is related to the endogenous proline content and in −Mo tissue where the proline reaches very high levels relative to other amino acids it is often difficult to inhibit the enzyme induction even by high levels of L-azetidine-2-carboxylic acid. In +Mo sterile cultures with ammonium sulphate where the proline content is almost as high, in absolute terms, as in the −Mo, but relative to other amino acids is not high, L-azetidine-2-carboxylic acid, even at moderate levels, inhibits the enzyme induction.

The specific effect of L-azetidine-2-carboxylic acid as an analogue of an essential amino acid (L-proline) for protein synthesis seems to us to be the best evidence so far for assuming that the induction of nitrate reductase in plants by nitrate or by the constituent metal molybdenum both involve protein synthesis *de novo* particularly as proline has been implicated as an important factor, if not the only one, in determining tertiary structure of proteins."

In cooperation with Afridi he had also found that methyltryptophane isomers, *p*-fluorophenylalanine, canavanine and β-thienyl-analine had virtually no effect on the inhibition of induction of nitrate reductase, but some others(L-2-thiolhistidine, L-dihydroxyphenylalanine and D,L-α-methyl-glutamic acid) had also appeared markedly inhibitory to induction.

Section 3

Interaction of Nitrogen Metabolism with External Factors

Section 3a

Relations between Plant Growth Regulators and Nitrogen Metabolism

L. C. LUCKWILL

Long Ashton Research Station,
University of Bristol, England

INTRODUCTION

The purpose of this review is to draw attention to some of the links which have been established between nitrogen metabolism and growth regulators. The latter term is used in its broadest sense to include not only the endogenous hormones of the plant but also some of the many synthetic growth regulators used in agricultural and horticultural practice as herbicides. Although many such links have been established, it is not always possible to pin-point with any degree of certainty the precise mechanisms involved; nor is this likely to be possible until the complex chemical pathways of nitrogen metabolism in the plant are themselves understood in more detail.

The relations between nitrogen metabolism and endogenous hormones are reciprocal. Not only do these hormones control certain phases of protein synthesis and degradation, but two of the three main classes of hormones, the auxins and the cytokinins, are themselves nitrogen-containing compounds whose production is inevitably linked with the nitrogen metabolism of the plant. The principal auxin in plants, indole-3-acetic acid, is a deamination product of tryptophan, produced probably *via* indolepyruvic acid and indoleacetaldehyde, and its production is associated particularly with regions of active protein synthesis at the meristems. Cytokinins are substituted amino-purines, though only one naturally occurring member of this group has so far been identified chemically. This is zeatin, a hydroxy derivative of dimethylallylaminopurine. Compounds of this type, which play a key role in the regulation of protein synthesis are thought to be produced in the root system and translocated in the xylem sap.

The main effects of growth regulators on nitrogen metabolism will be considered under three main headings. The first concerns the oxidation of nitrogen in the soil through the medium of nitrifying bacteria; the second phase is the enzymic reduction of nitrate in the plant, whilst the third phase covers the synthesis of proteins and their eventual degradation during senescence to simpler and more soluble forms of nitrogen.

OXIDATION OF NITROGEN IN THE SOIL

The recent extensive use of soil-applied herbicides of the substituted triazine and urea types which can persist under field conditions for many months, or even years, has prompted investigations into the possible effects of these herbicides on the nitrifying bacteria that play an essential role in the maintenance of soil fertility. Caseley and Luckwill (1965) at Long Ashton investigated the effects of a range of triazines and substituted ureas on soil nitrification, using a soil percolation technique. They also studied the effects of the herbicides on the growth in culture of *Nitrosomonas europaeus*, *Nitrobacter agile* and *Azotobacter vinelandii*. Of the herbicides tested, only monuron (*N'*-*p*-chlorophenyl-*NN*-dimethylurea) was found to have an inhibiting effect on soil nitrification at the rates commonly used in agricultural practice. This confirmed the earlier observations of Gamble *et al.* (1952) and of Quastel and Scholefield (1953). *Nitrosomonas*, which converts ammonia nitrogen to nitrite, was found to be more tolerant of monuron than was *Nitrobacter*, which oxidizes nitrite to nitrate, with the result that monuron-treated soil showed an accumulation of abnormal amounts of nitrite. Monuron in concentrations of 100 ppm or more in the soil solution was also found to inhibit the fixation of atmospheric nitrogen by *Azotobacter*. Monuron is highly adsorbed on soil colloids in the uppermost horizons so that, except in very sandy soils, this disturbance of the normal pattern of nitrogen metabolism is likely in practice to be strictly localized. Eventually, as the herbicide is broken down by microbial action, the soil will finally recover its full nitrifying capacity.

REDUCTION OF NITROGEN IN THE PLANT

In plants abundantly supplied with nitrate, the rate at which this nitrate can be assimilated into simple organic forms and eventually into protein, will depend largely on the activity of the nitrate and nitrite reductase systems in the plant and, under certain conditions, these may well become limiting factors controlling the rate of dry weight increase. It is known that nitrate reductase is an inducible enzyme and that induction of activity is dependent on the presence of both nitrate and molybdenum (Afridi and Hewitt, 1964). The induction of the enzyme, though not its *in vitro* activity, has been shown to be inhibited by a wide range of antimetabolites known to block protein metabolism, and to be stimulated by serine (Afridi and Hewitt, 1965).

EFFECTS OF SUB-LETHAL DOSES OF HERBICIDES

There is good evidence that nitrate assimilation and utilization in certain crops can be enhanced by the application of sub-lethal doses of certain herbicides, and, although the precise mechanism has, in most instances, not been thoroughly investigated, a stimulatory effect on the induction of nitrate reductase or on the activity of the enzyme would seem to be indicated. These effects are most frequently seen in crops possessing a high degree of tolerance to the herbicide. Tolerance implies the existence of detoxication mechanisms that can quickly metabolize the herbicide as it enters the tissues and maintain the concentration well below the level at which symptoms of phytotoxicity occur. In the absence of such detoxication processes, any stimulatory effects on growth or nitrate assimilation are likely to be masked by the phytotoxic effects of the herbicide, unless the latter has been applied at a very low rate.

Effects of DNOC on cereals. Spraying a crop of winter cereals with DNOC (2,4-dinitro-*o*-cresol) for weed control generally results in an initial retardation of growth lasting about 2 months. When this initial phytotoxic effect wears off, subsequent growth is stronger and the plants have an enhanced chlorophyll and nitrogen content as compared with unsprayed plants. This stronger growth is usually accompanied by increased yield and delayed ripening of the grain. Bruinsma (1963) working with winter rye (*Secale cereale* L.) has shown that this increased yield cannot be attributed to the absence of weed competition, because it was obtained even when the experiment was conducted in the absence of weeds. He has also calculated that the nitrogen content of the DNOC is quite inadequate to account for the magnitude of the effects observed. His conclusion is that DNOC affects growth and yield by enabling the crop to make better use of the nitrogen present in the plant and the soil.

Effects of triazine herbicides. The stimulating effects of simazine (2-chloro-4,6-bisethylamino-1,3,5-triazine) and atrazine (2-chloro-4-ethylamino-6-isopropylamino-1,3,5-triazine) on the nitrogen metabolism of crops that are tolerant of these herbicides is frequently noted in horticultural experiments. The best authenticated example is probably that of simazine on maize. Maize is tolerant to simazine, at the rates normally used for weed control, by virtue of its ability to metabolize the herbicide to a non-phytotoxic derivative, 2-hydroxy-simazine. Ries and Gast (1964) grew maize in sand culture with three levels of nitrogen, supplied as ammonium nitrate, and three levels of simazine. Simazine, at all concentrations, greatly enhanced the growth and nitrogen content of the maize. At the lowest level, the effect of 3 mg simazine was equivalent to that of an additional 570 mg of nitrogen as ammonium nitrate. Similar information is provided by Karnatz (1965) who estimated that 10 kg/ha simazine was equivalent to 200 to 250 kg/ha of nitrogen in its effect on the nitrogen content of apple leaves. In maize it has been confirmed that this effect is due to unmetabolized simazine and not to 2-hydroxy-simazine.

In more recent work, Ries *et al.* (private communication) have shown striking increases in nitrate reductase activity in maize, cucumber, lettuce and rye plants following treatment with low doses of simazine. Ries has also observed that these effects are manifested only in plants growing under sub-optimal conditions of temperature and nitrate nitrogen levels. This raises the interesting possibility of using simazine, or some non-phytotoxic analogue, not as a herbicide but as a growth stimulator to increase the efficiency of plants growing in sub-optimal environments. Gramlich *et al.* (1965) have shown a similar stimulating effect of atrazine on nitrate reductase activity in maize, measured 5–12 days after treatment. *In vitro*, atrazine had no effect on the activity of a cell-free nitrate reductase preparation from maize, from which it was deduced that, in the plant, the herbicide is stimulating the induction of the enzyme. An alternative explanation suggested by the work of Hewitt (unpublished) is that simazine may increase the stability of the enzyme rather than stimulate its induction. This idea is based on the observation that nitrate reductase persists longer after deprival of nitrate in simazine-treated than in untreated plants.

In fruit crops there are many reports of growth stimulation, accompanied by increased nitrogen content of leaves and increased yield following herbicidal applications of simazine. Luckwill and Caseley (1966) have reported yield increases of up to 27% over a three-year period as a result of treating black currants with an annual application of 8 lb/acre simazine. On apples, a crop that can rapidly detoxify simazine (Luckwill and Caseley, 1966), very striking increases in vegetative growth can follow the use of this herbicide on the soil. Ries *et al.* (1963) found that the treatment of young apple trees of three different varieties with soil applications of simazine at 4 lb/acre almost doubled the amount of the terminal growth, under conditions where ¼ lb ammonium nitrate on the tree had no effect. Similar stimulating effects of simazine and atrazine on vegetative growth and on nitrogen content of the leaf have been reported for apples and grapes by Gast and Grob (1964), for peach by Ries *et al.* (1963), for Shamouti orange by Goren and Monselise (1966) and for potato and *Convolvulus arvensis* by Ries and Gast (1964). There is no doubt that these effects are real and that, for the fruit grower they pose special practical problems concerned with choice of rootstock and with the level and time of application of nitrogenous fertilizers. These problems are particularly acute in apples, a crop in which simazine is very widely used and in which excessive vegetative vigour is often accompanied by a reduction in flower initiation. It may be reasonably postulated that in fruit crops, as in maize, the stimulating effect of the herbicide is in the induction or stabilization of nitrate reductase, but more detailed work is needed to confirm this.

Effects of phenoxy-acids. 2,4-dichlorophenoxyacetic acid (2,4-D) is another growth regulator reported to have effects on the nitrate reductase system. Beevers *et al.* (1963), working with cell free extracts of maize and cucumber showed an increased level of nitrate reductase activity in maize, and a reduced level in cucumber, as a result of treatment of the plants with 2,4-D at concentrations of 10 and 100 ppm. In view of the rather high concentrations of 2,4-D used, the apparent differences in activity on the two species might well be due to differences in the level of free 2,4-D in the tissues, since the resistant maize will almost certainly metabolize the herbicide much faster than the sensitive cucumber. 2,4-D is known to influence the endogenous levels of sulphydryl in certain plants and Beevers *et al.* suggest that, since nitrate reductase is a sulphydryl-containing enzyme, the effects of 2,4-D might be explained on the basis of a modification of enzyme structure through intramolecular changes in protein sulphydryl groups. As an alternative mechanism they suggest that the effects of 2,4-D on the activity of nitrate reductase might be mediated through its effects on nucleotide metabolism, as demonstrated by Key *et al.* (1960). Further work is needed to differentiate between these and other possibilities. In view of the activity of 2,4-D it would also be of interest to determine whether the native auxin, IAA, might play some part in controlling the activity of nitrate reductase either *in vitro* or *in vivo*. Meanwhile, it is interesting to speculate on the possible relationship between this work and the field experiments of Wort (1964) who has convincingly demonstrated significant increases in yield of sugar beet, potatoes and other crops as a result of spraying with sub-lethal concentrations of 2,4-D, either alone or in combination with certain micronutrient elements.

Effects of uracil derivatives. Bromacil (5-bromo-3-*sec.*-butyl-6-methyluracil) and isocil (5-bromo-3-isopropyl-6-methyluracil) are herbicides which, like simazine, are thought to exert their primary phytotoxic effects through an inhibition of photosynthesis. Unlike simazine, however, both these uracil derivatives, in concentrations ranging from 1 to 10 μg/ml have been shown to be potent inhibitors of the induction of nitrate reductase in leaf discs of radish and cauliflower (Hewitt and Notton, 1966). The mechanism of action is not yet known, though the authors suggest that nitrate reductase, like nitrite reductase, may be located in the chloroplasts and that a common site, if not a common mechanism, for the inhibition of photosynthesis and the induction of nitrate reductase may be involved. However, this suggestion is not supported by the more recent work of Schrader *et al.*, (1967) with chloramphenicol, which is known to be specific for the inhibition of protein synthesis in chloroplasts. They found that chloramphenicol depressed nitrite reductase but actually stimulated nitrate reductase activity in green leaf tissue, a fact which they interpret as consistent with the suggestion that nitrite reductase is located in the chloroplasts, and nitrate reductase in the cytoplasm. Moreover, the observation that in many plants the reduction of nitrate and nitrite takes place in the root system (see Sections 1e, 3b and 3d) would seem to rule out any possibility of a relationship between the photosynthetic and nitrate reducing systems.

Chlorate toxicity. A curious link between nitrate reductase and herbicidal action relates to the mechanism of action of sodium chlorate as a weed-killer, reported by Liljeström and Åberg (1966). They grew wheat plants with a greatly decreased nitrate reductase content by using urea as the sole source of nitrogen, and found that such plants showed greatly enhanced tolerance to chlorate. When the enzyme was inhibited by cyanate, chlorate toxicity was completely eliminated. They conclude that chlorate in the plant is reduced to chlorite by the same enzyme system that reduces nitrate, and that chlorite is the toxic agent responsible for the death of the plant. Although a relationship has undoubtedly been established, further investigation would seem desirable to confirm their claims that one and the same enzyme is involved in the reduction of chlorate and nitrate, and also to determine to what extent chlorate might replace nitrate as an inducer of the enzyme.

PROTEIN SYNTHESIS AND DEGRADATION

HERBICIDES THAT AFFECT PROTEIN SYNTHESIS

Carbamates. Carbamate herbicides such as propham (isopropyl *N*-phenyl carbamate) and chlorpropham (isopropyl *N*-(3-chlorophenyl)carbamate) have been in common use for some years for controlling certain annual weeds in the seedling stage, and more recently other members of this group such as "swep" (methyl-*N*-(3,4-dichlorophenyl)carbamate) and barban (4-chlorobut-2-ynyl *N*-(3-chlorophenyl)carbamate) have found employment in horticultural practice. In susceptible plants, these carbamate herbicides can produce many diverse effects on physiological processes. Propham and chlorpropham have long been known as respiratory poisons, inhibiting the dehydrogenases of the four carbon acid cycle. They are also mitotic poisons. More recent information indicates that their primary mode of action may be the blocking of amino acid incorporation into protein. Mann *et al.* (1965) found that all four of the carbamates listed above would inhibit the incorporation of ^{14}C-labelled methionine into radioactive polymers, consisting largely of protein, and that this inhibition could be detected within 30 minutes of treatment.

Toxicity of amitrole. Amitrole (3-amino-1,2,4-triazole) is a widely used non-selective herbicide, the most characteristic feature of which is the induction of complete chlorosis of all tissue developing after treatment. This is due to a complete failure of the tissues to develop plastids. This compound affects so many aspects of plant metabolism that it is difficult to define its primary mode of action. For instance, amitrole can form metal chelates which, by locking up essential micronutrients, might lead to a reduction in the activity of many different enzymes. It is also capable of forming relatively stable complexes with glycine and alanine (Naylor, 1964) and probably also with serine (Carter and Naylor, 1961), which might be expected to restrict the synthesis of both purines and proteins. However, the most specific effect of amitrole seems to be in blocking the formation of L-histidine, required for the synthesis of proteins and nucleic acids. Indirect evidence of this comes from the work of Hilton (1960), Jackson (1961) and Casselton (1964), all of whom have demonstrated that the toxic effects of amitrole can be overcome, at least partially, by supplementary treatment with histidine. Moreover, the chlorotic tissue of amitrole-treated plants has been shown to be deficient in protein, but abnormally rich in amino acids and free ammonia (McWhorter, 1963). In amitrole-treated yeast (Hilton and Kearney, 1964) and unicellular algae (Siegel and Gentile, 1966) the accumulation in the external solution of imidazolylglycerol, a known precursor of L-histidine, has been observed.

Dalapon. When sugar beets, which are resistant to dalapon (2,2-dichloropropionic acid), and *Setaria glauca*—a grass which is susceptible—are treated with this herbicide, both species show striking increases in free amino acids and amides in the tissues six days after treatment. Since there is no appreciable change in total nitrogen during this time, this soluble nitrogen is assumed to come from the degradation of protein (Andersen *et al.*, 1962), though it is not clear at present whether dalapon is stimulating degradation or inhibiting synthesis. In the resistant sugar beets, probably as the result of the metabolism or inactivation of the herbicide in the tissues, protein metabolism returns to normal before any permanent damage is done. This does not happen in the susceptible *Setaria*.

Phenoxyacids. There is a great deal of published work, much of it conflicting, on the effect of 2,4-D and other phenoxyacid herbicides on protein content and distribution in plants. Freiberg and Clark (1952) found a decrease in protein nitrogen in the leaves of treated soya bean plants, but an increase in the stems and roots. In addition, all organs of the plant showed an increase in soluble organic nitrogen. By contrast, Woodbridge and Kamal (1962), working on Bartlett pear, found that 2,4-D treatments generally increased protein and total nitrogen in leaves, but decreased them in stem tissue. The general conclusion to be drawn from published work would seem to be that 2,4-D frequently affects the distribution of protein in the plant and sometimes, as might be expected from its stimulating effect on nitrate reductase, it will increase the total content of organic nitrogen.

EFFECTS OF ENDOGENOUS HORMONES

Senescence of leaves. The senescence of leaves, the visual symptom of which is a progressive loss of chlorophyll, is accompanied by a steady decline in the ability of the tissue to synthesize RNA and protein, and by the liberation of soluble nitrogen compounds that are either translocated to parts of the plant still in active growth or, in woody species, move back into the stems and roots for storage. There is now abundant evidence that this phase of nitrogen metabolism is under the direct control of endogenous hormones produced in other parts of the plant. Experimentally, senescence can be induced readily by detaching the leaf from the plant or by using leaf discs floating on water or supported on moist filter paper. The senescence of such detached leaves or leaf discs is delayed by treatment with auxins, gibberellins or cytokinins, depending on the species and the stage of growth. *Xanthium* leaves respond only to cytokinins; those of *Rumex obtusifolius* only to gibberellins; and those of *Prunus* only to auxins; but in some species, such as banana, all three types of hormone are active in delaying senescence (Osborne, 1965; Whyte and Luckwill, 1966). A synthetic kinin, benzyladenine, has been used commercially to delay senescence and thus maintain the freshness of leafy vegetables such as spinach and lettuce after harvest. A few substances are known that will actively promote senescence. Ethylene will induce leaf yellowing and abscission in many species and Beevers (1966) has recently shown that the growth retardant *N*-dimethylaminosuccinamic acid (B.9) will promote the senescence of leaf discs of *Tropaeolum majus*, an effect which can be counteracted by the simultaneous application of either kinetin or gibberellic acid. Higher concentrations of *N*-dimethylaminosuccinamic acid and other growth retardants, however, will retard the senescence of leaf-discs of *Rumex obtusifolius* (Harada, 1966).

Osborne and Hallaway (1964) have studied the effects of the auxin 2,4-D in retarding senescence in discs cut from autumn leaves of cherry. The ability of these leaves to synthesize protein was studied by feeding the leaf discs with ^{14}C-labelled leucine, and subsequently isolating and counting the protein. In the presence of 2,4-D protein synthesis was maintained at the orignal level, or even slightly enhanced, whereas in the controls it continued to decline, falling by as much as 60 % over a period of seven days. Experiments with leaf discs of *Xanthium* (Osborne, 1965) show that kinetin not only prevents the decline—but actively stimulates the synthesis of both RNA and protein as shown by the incorporation of labelled precursors. Puromycin, which is known to interfere with the condensation of amino acids on to the nucleoprotein template, completely inhibits this action of kinetin. Of special interest is the fact that when DNA-dependent RNA synthesis is partially blocked by actinomycin D, an antibiotic which complexes with the guanine base of DNA, then the stimulatory action of kinetin on protein synthesis is roughly proportional to the amount of DNA-dependent RNA synthesis still occurring in the tissue. From evidence of this kind, Osborne proposes that the action of kinetin in preventing senescence could be mediated through an effect on DNA-dependent RNA synthesis, possibly at the stage of messenger RNA synthesis.

Other hormone effects. An essentially similar mode of action has been suggested for gibberellin which, as shown by the work of Varner and Chandra (1964), induces the formation of the enzyme α-amylase in the aleurone layer of the barley grain. The induction of the enzyme by gibberellin is blocked by all the usual inhibitors of protein synthesis and, more significantly, also by actinomycin D. As with kinetin, it would appear that the site of action of gibberellin may lie very close to the genes themselves, in the control of messenger RNA synthesis.

More recently Nooden and Thimann (1966) have shown that even that most classical example of plant hormone action—auxin-induced cell extension—is correlated with, and possibly dependent on, protein synthesis. As with gibberellin, protein inhibitors and actinomycin D are active in blocking both protein synthesis and cell extension. Although protein synthesis now appears to be a critical factor in auxin-induced cell extension, it is not at present clear whether structural protein is involved or whether the auxin is controlling the induction of an enzyme affecting the plasticity of the cell wall.

Abscission is another auxin-controlled process recently found to be associated with RNA and protein synthesis (Abeles and Holm, 1966). Abscission in bean leaves was induced with ethylene, and protein and RNA synthesis in the region of the abscission zone was demonstrated by the use of labelled precursors. Carbon dioxide, which inhibits ethylene-induced abscission, also reduces RNA and protein synthesis, and both processes are blocked by actinomycin D. It would be interesting to extend these studies to cover the effect of auxin which in very low concentrations promotes abscission but at higher concentrations retards it.

Mobilization of nitrogen compounds. No review of the relations between nitrogen metabolism and plant hormones would be complete without some reference to the effects of cytokinins in mobilizing the soluble organic nitrogen constituents of the plant. This effect was first dramatically demonstrated by Mothes and Kulajewa (1959) who, by treating one side of a leaf with kinetin were able to demonstrate a movement of labelled amino-acids from the other untreated half into the region of high kinetin concentration. The mechanism of this effect is still obscure, but since non-protein amino acids such as aminoisobutyric acid will move readily towards a kinin source, it is unlikely to be merely a source/sink effect. Leopold and Kawase (1964), by treating the primary leaves of bean plants with the synthetic cytokinin, benzyladenine, were able to induce the premature senescence of the younger trifoliate leaves. This is a reversal of the normal state of affairs where the mobilization of soluble nitrogen by the cytokinin-rich younger leaves on the plant leads to senescence of the older leaves. The phenomenon of hormone-directed transport of nutrients is not confined to cytokinins and nitrogenous materials: auxins produce much the same effect, and carbohydrates, as well as amino acids, can be mobilized. This was shown as long ago as 1937 by Mitchell and Martin, who found that the application of auxin to the apices of decapitated pea plants led to the accumulation of soluble carbohydrates and nitrogenous material in the stumps. This phenomenon has since been confirmed in the more critical experiments of Booth *et al.* (1962) and also by Davies and Wareing (1965). The original concept of hormones as "chemical messengers" is in fact now being replaced by a new concept of "chemical policemen"—directing the flow of nutrients in the plant to sites where they are required for protein synthesis.

ACKNOWLEDGEMENTS

The author is grateful to his colleague, Dr E. J. Hewitt, for his critical comments on the manuscript and to Dr S. K. Ries, Department of Horticulture, Michigan State University, for permission to quote unpublished work on the influence of simazine on nitrate reductase.

REFERENCES

Abeles, F. B. and Holm, R. E. (1966). *Pl. Physiol.* **41**, 1337.
Afridi, M. M. R. K. and Hewitt, E. J. (1964). *J. exp. Bot.* **15**, 251.
Afridi, M. M. R. K. and Hewitt, E. J. (1965). *J. exp. Bot.* **16**, 628.
Andersen, R. N., Behrens, R. and Linck, A. J. (1962). *Weeds* **10**, 4.
Beevers, L. (1966). *Pl. Physiol.* **41**, suppl. lxxiv.
Beevers, L., Peterson, D. M., Sannon, J. C. and Hageman, R. H. (1963). *Pl. Physiol.* **38**, 675.
Booth, A., Moorby, J., Davies, C. R., Jones, H. and Wareing, P. F. (1962). *Nature, Lond.* **194**, 204.
Bruinisma, J. (1963). *Pl. Soil*, **18**, 1.
Carter, M. C. and Naylor, A. W. (1961). *Physiologia Pl.* **14**, 62.

Caseley, J. C. and Luckwill, L. C. (1965). *Rep. Long Ashton Res. Stn for* 1964, 78.
Casselton, P. J. (1964). *Nature, Lond.* **204**, 93.
Davies, C. R. and Wareing, P. F. (1965). *Planta* **65**, 139.
Freiberg, S. R. and Clark, M. E. (1952). *Bot. Gaz.* **113**, 322.
Gamble, S. J. R., Mayhew, C. J. and Chappell, W. E. (1952). *Soil Sci.* **74**, 347.
Gast, A. and Grob, J. (1964). *Proc. 7th Brit. Weed Control Conf.* 217.
Goren, R. and Monselise, S. P. (1966). *Weeds* **14**, 141.
Gramlich, J. V., Davies, D. E. and Funderburk, H. H. (1965). *Proc. 18th sth. Weed Control Conf.* 1965, 611.
Harada, H. (1966). *Pl. Cell Physiol. Tokyo*, **7**, 701.
Hewitt, E. J. and Notton, B. A. (1966). *Biochem. J.* **101**, 39c.
Hilton, J. L. (1960). *Weeds* **8**, 392.
Hilton, J. L. and Kearney, P. C. (1964). *Abstr.* 1964 *Mtg. Weed Soc. Am.* 75.
Jackson, W. T. (1961). *Weeds* **9**, 437.
Karnatz, H. (1965). *Z. PflKrankh. PflPath. PflSchutz* Sonderheft 3, 149.
Key, J. L., Hanson, J. B. and Bils, R. F. (1960). *Pl. Physiol.* **35**, 177.
Leopold, A. C. and Kawase, M. (1964). *Am. J. Bot.* **51**, 294.
Liljeström, S. and Åberg, B. (1966). *K. Lantbr. Högsk. Annlr.* **32**, 93.
Luckwill, L. C. and Caseley, J. C. (1966). *In* "Herbicides in British Fruit Growing" (J. D. Fryer, ed.) p. 81. Blackwell, London,
Mann, J. D., Jordan, L. S. and Day, B. E. (1965). *Weeds*, **13**, 63.
McWhorter, C. G. (1963). *Physiologia Pl.* **16**, 31.
Mitchell, J. W. and Martin, W. E. (1937). *Bot. Gaz.* **99**, 171.
Mothes, K. and Kulajewa, O. (1959). *Flora, Jena* **147**, 445.
Naylor, A. W. (1964). *J. agric. Fd Chem.* **12**, 21.
Nooden, L. D. and Thimann, K. V. (1966). *Plant Physiol.* **41**, 157.
Osborne, D. J. (1965). *J. Sci. Fd Agric.* **16**, 1.
Osborne, D. J. and Hallaway, M. (1964). *New Phytol.* **63**, 334.
Quastel, J. H. and Scholefield, P. G. (1953). *Appl. Microbiol.* **1**, 282.
Ries, S. K. and Gast, A. (1964). *Abstr.* 1964 *Mtg. Weed Soc. Am.* 72.
Ries, S. K., Larsen, R. P. and Kenworthy, A. L. (1963). *Weeds*, **11**, 270.
Schrader, L. E., Beevers, L. and Hageman, R. H. (1967). *Biochem. biophys. Res. Commun.* **26**, 14.
Siegel, J. N. and Gentile, A. C. (1966). *Pl. Physiol.* **41**, 670.
Varner, J. E. and Chandra, G. R. (1964). *Proc. natn. Acad. Sci. U.S.A.* **52**, 100.
Whyte, P. and Luckwill, L. C. (1966). *Nature, Lond.* **210**, 1360.
Woodbridge, C. G. and Kamal, A. L. (1962). *Proc. Am. Soc. hort. Sci.* **81**, 116.
Wort, D. J. (1964). *In* "The physiology and biochemistry of herbicides". (L. J. Audus, ed.) Chap. 10, Academic Press, London and New York.

Section 3b

Deviations in Nitrogen Metabolism Associated with Viruses

ROY MARKHAM*

Agricultural Research Council,
Virus Research Unit, Cambridge, England

The effects of virus infection on the general pattern of nitrogen metabolism of the host plant are extremely difficult to interpret, even though it is now realized that viruses, unlike other pathogens, act by means of a mechanism whereby new proteins are the primary products of the processes of virus replication. The reasons for this difficulty can be seen in the enormous variability of the symptom patterns produced in different virus infections. Death of the whole organism, though known to occur, is uncommon, although the even more extreme case, in which the death of localized areas of tissue prevents systemic spread of the virus, is well known, and is thought to be in the nature of a defence reaction. At the other extreme, virus infection may cause little or no outward signs even though the quantity of virus actually made may be quite large. This is to be seen particularly in infections of potato plants with the potato "X" virus. In this instance, wild strains of virus have been selected by the process involved in the "roguing" of obviously diseased plants, and the remaining plants not only show no symptom, but the gross effects of infection can only be demonstrated by fairly complex field experiments (Smith and Markham, 1945). In between these two extremes all manner of effects occur, such as the production of mosaics, in which chlorophyll production is reduced partially or, in some cases, almost completely, systemic necrosis, distortion, tumour production and, occasionally, such severe damage to vascular tissues that the translocation of metabolites, particularly carbohydrates, is severely impeded. This last condition is to be found in such infections as those caused by potato leaf-roll virus and the sugar beet yellows viruses, which are characterized by loss of chlorophyll and accumulation of starch in the leaf tissues which become stiff and may also develop colouring due to the production of anthocyanin pigments. Sometimes symptoms may mimic severe hormone damage. Other symptom patterns include the reversion of plant characters towards a "wild" type, and even the failure of specialized

* Now Director of the John Innes Institute, Norwich, and Professor of Cell Biology, University of East Anglia.

organs to differentiate. The latter is found in cases of "phyllody", which is not uncommon in clovers among many other plants, and in which the flowers develop as bunches of leaves, instead of differentiating normally.

In view of the diversity of patterns produced by virus infection, it is not altogether surprising that the examination of the metabolic pools in diseased plants has not been entirely rewarding. In the early stages of virus infection the production of virus protein may, indeed, be a major metabolic activity, the quantity of virus protein formed following infection by viruses such as the tobacco mosaic and turnip yellow mosaic viruses being as much as 2–3% of the dry weight of the host plant. Examination of the metabolic pools has, however, not yielded much information, nor indeed would it be expected to, though considerable work has been done on plants infected with wild strains of viruses such as the potato "X" virus (Miczynski, 1959). It has also been suggested (Wildman *et al.*, 1949) that virus synthesis takes place at the expense of a high molecular weight protein (18S protein or Fraction 1 protein) found in sap extracts (Fig. 1), but an intensive survey carried out in my laboratory (M. W. Rees and M. W. Johnson, unpublished) failed to give any indication that this was so. In these experiments, sap specimens were examined at regular intervals over a period of many weeks by means of the analytical ultracentrifuge.

There are, however, two outstanding instances of abnormalities in the metabolism of nitrogenous compounds in virus-diseased plants. The first of these is the accumulation of pipecolic acid (piperidine 2-carboxylic acid) in peach plants having Western "X" virus disease. This imino acid is detectable in the young leaf-tissue of healthy peaches, but accumulates in quantity in the older leaves of diseased plants (Diener and Dekker, 1954) This accumulation appears to be due to this specific virus disease, but is also found in peaches suffering from arsenic toxicity. Pipecolic acid has also been reported to be present in small amounts in other virus diseased plants (Bozarth and Diener, 1963). The other instance was uncovered during a study of the cation content of a number of plant viruses in an attempt to discover how the phosphate groups of the nucleic acid were neutralized (Johnson and Markham, 1962). In this investigation quite appreciable amounts of a triamine, hitherto unrecognized in biological systems, were found as a constituent of several viruses, including turnip yellow mosaic virus, but this compound was not found in uninfected plants. This triamine is the symmetrical 1 : 7 diamino-4-azaheptane, and its particular interest is that the evidence suggests that it is a product coded for by the nucleic acid of the viruses. Its structure, of course, is one that is extremely likely to bind specifically to nucleic acids, and, indeed, it has a stabilizing effect on viral nucleic acid (Johnson and Hills, 1963).

The question as to how much information is carried by a virus, and how little is really essential for its economy is one that has attracted considerable attention in recent times. This problem is not one that can be tackled at all readily with the available plant-plant virus systems, but much can be deduced indirectly. For example in the case of the large T-even bacteriophages of *Escherichia coli*, infection causes the production *de novo* of a number of new enzymes implicated in the replication of the virus (Bello *et al.*, 1961; Aposhian and Kornberg, 1962; Mathews *et al.*, 1964). The overall effect of the production of these enzymes is to inhibit the formation of bacterial deoxyribonucleic acid, and to induce the production of virus constituents. Plant viruses, especially the small ones, are much less complex than these giant bacteriophages and so cannot contain as much coded information, and thus their potentialities are not so great. However, one can infer from the analytical information available at present that about 3000 nucleotides are absolutely necessary for an autonomous virus. The broad bean mottle and cowpea chlorotic mottle viruses have about this complexity and would seem to represent the smallest "true" viruses, and they would appear to be able to code for five or six new proteins having molecular weights in the region of 18,000–25,000. These would, of course, include the viruses' own coat protein, and presumably a ribonucleic acid replicating enzyme. The remainder may well be kinases and probably the enzyme for triamine synthesis.

The actual amount of the enzymes produced would probably preclude their detection other than by virtue of their activity, and at present attention is being concentrated on the study of virus-induced DNA independent ribonucleic acid replicases (Sänger and Knight, 1963). Several claims have been made to have detected these enzymes (see Markham *et al.*, 1963, and Bovét, 1967).

One type of plant virus is known, the tobacco necrosis "satellite" virus, in which the amount of information stored in its nucleic acid is reduced below this apparent minimum, and presumably for this reason this virus is itself dependent for its multiplication upon a more complex virus, one of the tobacco necrosis viruses (Kassanis and Nixon, 1961). In the "satellite" virus, (Fig. 2), which itself may be regarded as a product of virus activity, the quantity of nucleic acid is such that it is only sufficient to code for about two proteins (Reichmann *et al.*, 1962). It has, indeed, been suggested that the coat protein of this virus is unusually complex and that the only protein coded for by the nucleic acid is the coat, but for various reasons this would appear to be improbable. In any case this is further confirmation that the ability to code for a number of proteins is an absolute necessity for an autonomous virus.

The "coat" itself poses an interesting problem. The average small virus' coat contains from 180 peptide molecules up to several thousand and it is

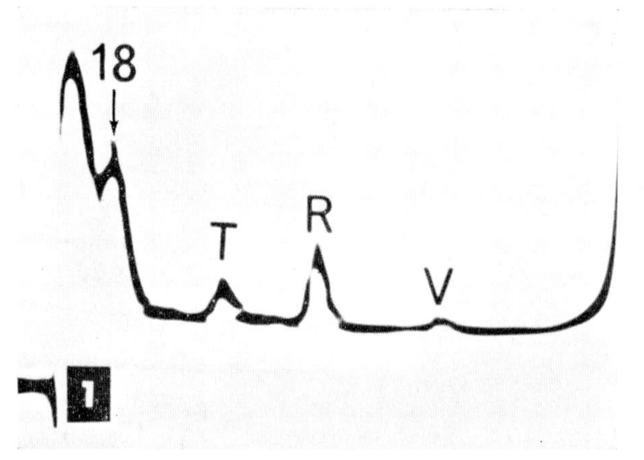

FIG. 1. Ultracentrifuge Schlieren diagram from a run on sap from pumpkin infected with Wild Cucumber Mosaic Virus. The peaks labelled are: 18, the 18S or Fraction 1 protein; T, empty virus shells; R, ribosomes; V, whole virus particles.

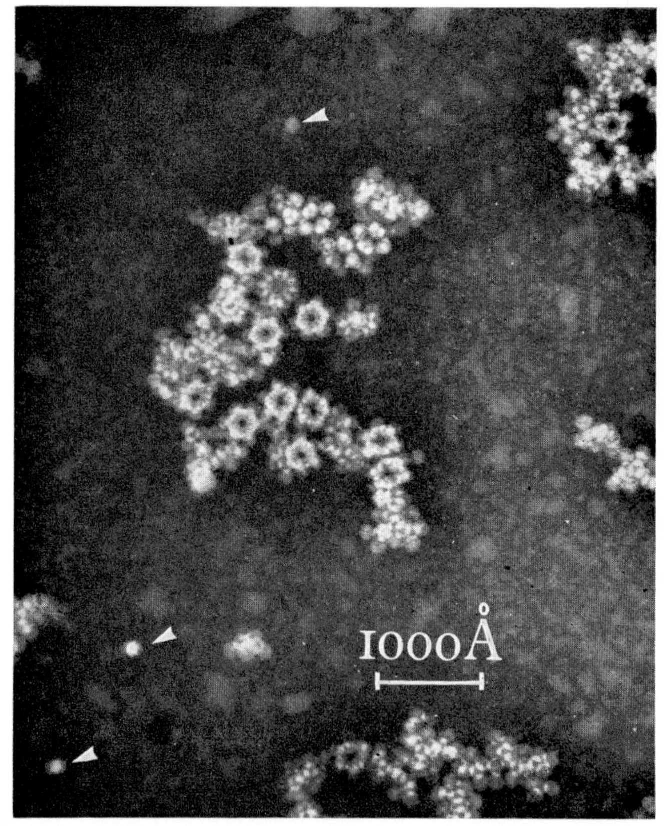

FIG. 2. Electron micrograph of the "Satellite" virus of Tobacco Necrosis. In this specimen almost all the virus particles are in the form of the semi-crystalline dodecamer. Single particles are indicated by arrows.

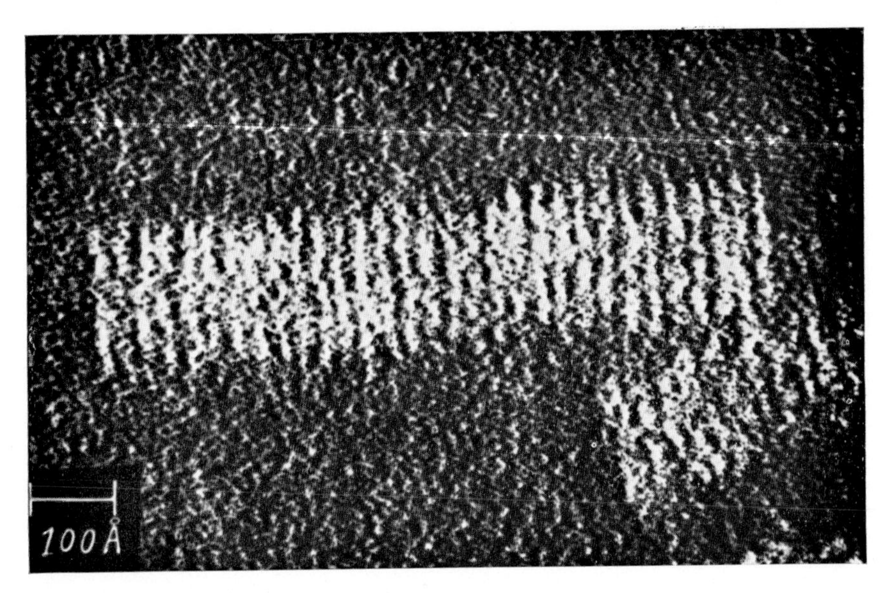

FIG. 3. Electron micrograph of the Stacked Disc form of the Tobacco Mosaic Virus.

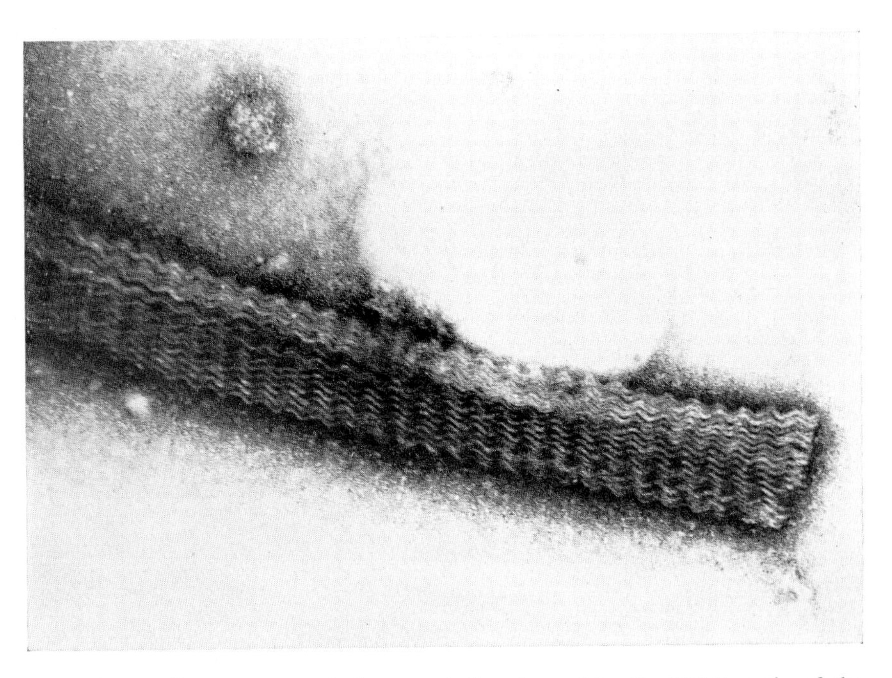

FIG. 4. Crystalline aggregate of the open helices formed by the PM 2 strain of the Tobacco Mosaic Virus.

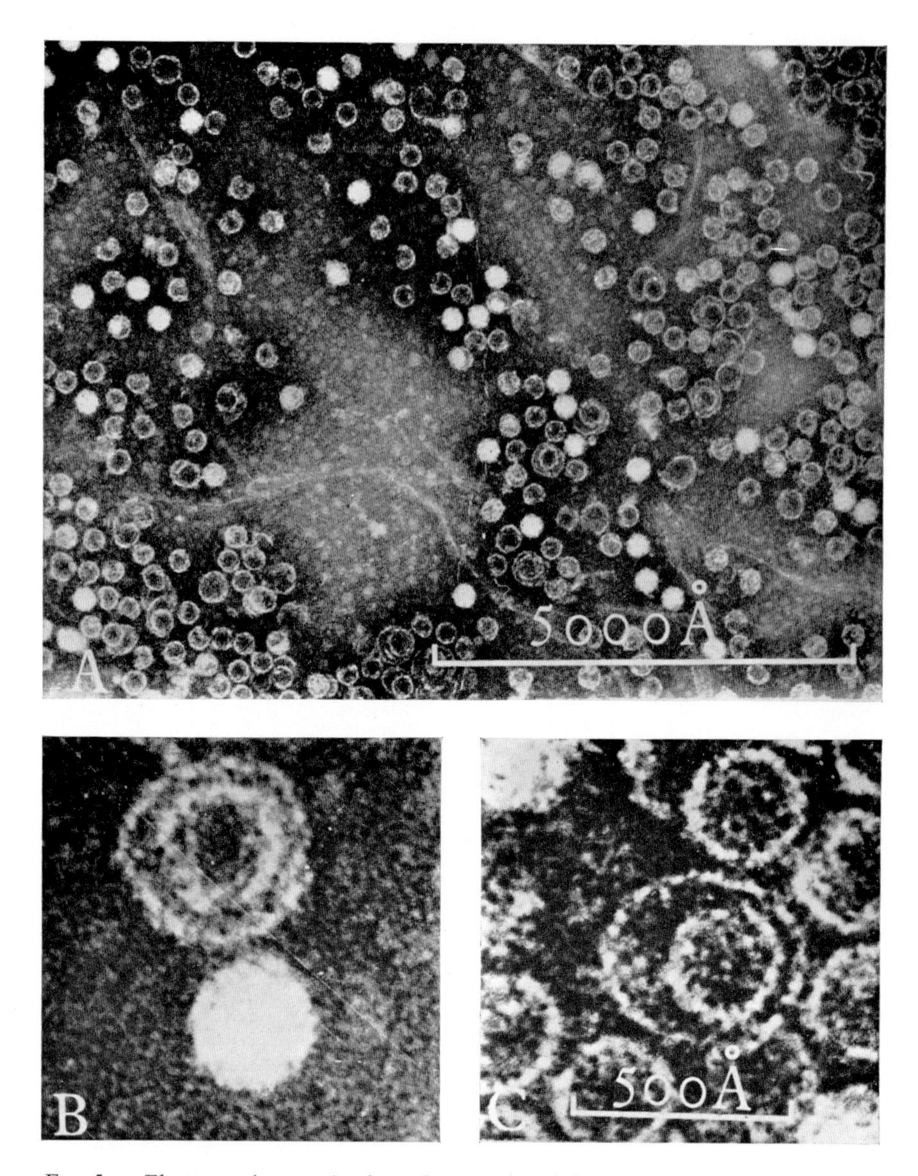

FIG. 5. A. Electron micrograph of sap from a plant infected with the Wild Cucumber Mosaic Virus, showing whole virus particles, empty shells and empty shells surrounded by large shells; B. *top*, a double shell, *bottom*, a virus particle; C. a double shell surrounded by empty shells.

FIG. 6. Flattened tubes isolated from sap of Chinese cabbage infected with a strain of Turnip Yellow Mosaic Virus.

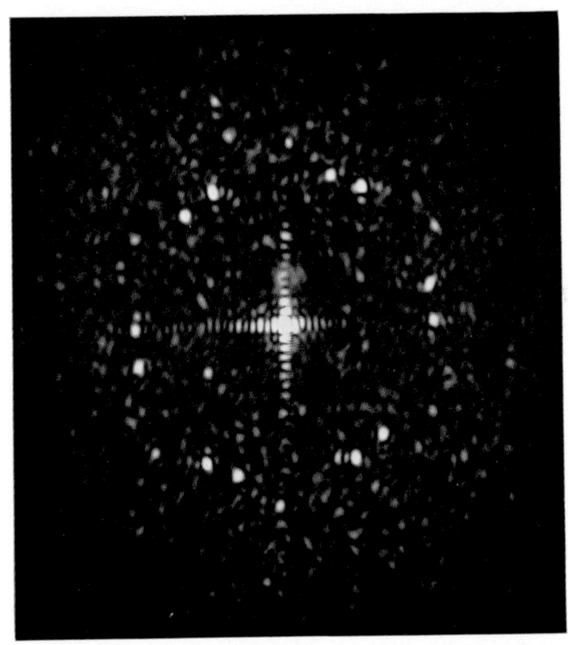

FIG. 7. Optical diffraction pattern from a flattened tube of the type shown in Fig. 6, showing that it is very perfect and formed from structures arranged hexagonally.

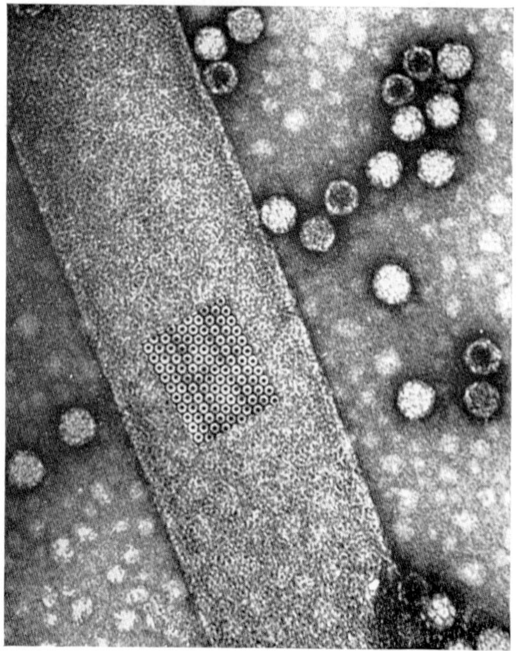

FIG. 8. A flattened tube of Turnip Yellow Mosaic protein surrounded by virus particles. Detail from both sides is visible and each obscures the structure of the other. Inserted is an averaged picture of one side of the tube obtained by optical integration (Markham *et al.* 1964) showing that the tube is made of hexagonal subunits packed in a hexagonal array.

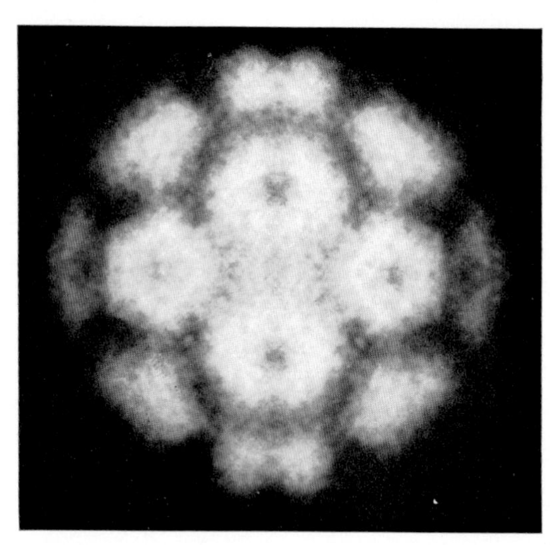

FIG. 9. An electron micrograph of Turnip Yellow Mosaic Virus, processed according to Bancroft *et al.* (1967) and showing that the surface is composed of pentagonal and hexagonal rings. The latter correspond to the substructures shown in Fig. 8 insert ($\times$2 million).

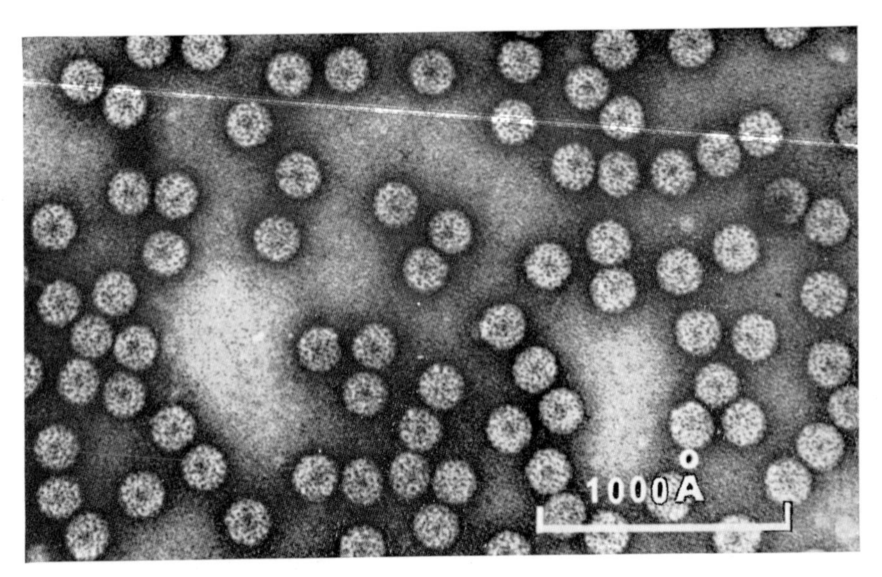

FIG. 10. Electron micrograph of the Cowpea Chlorotic Mottle Virus, showing the uniformity of the particles and their approximately spherical shape (cf. Fig. 9).

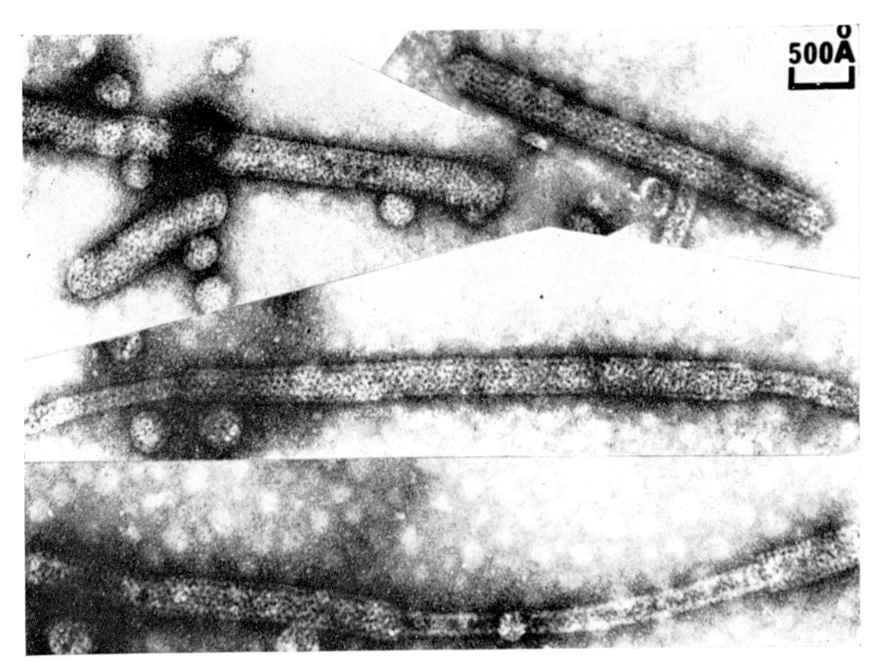

FIG. 11. Tubular structures formed from nuclease-treated Cowpea Chlorotic Mottle Virus. The spherical particles seen are also not virus, but are made from virus protein.

FIG. 12. "Bag" inclusion of a strain of Potato "Y" Virus from Mauritius.

quite obvious that the precise specification of a whole virus particle having this degree of complexity is quite beyond the coding ability of any normal virus nucleic acid. The solution that seems to have been adopted universally among the smaller plant viruses is to produce a small protein, containing some 150–220 amino acid residues, which folds to form a small, compact structure that has built-in polymerizing potentialities. This ability enables the protein to enclose the viral nucleic acid automatically, and so to protect it from the dangers of its normal environment. This would appear at first sight to be an exquisitely simple and practical solution to the problem. So it is, but it seems in many, and possibly in all, instances to have evolved at some expense, by inefficiency in the use of protein. The first real indication of this was the discovery (Markham and Smith, 1949) that, in plants infected by the turnip yellow mosaic virus, a substantial proportion of the virus material produced was incomplete, devoid of nucleic acid and non-infectious. Since then, many similar instances have been recognized. Even the tobacco mosaic virus causes over-production of its coat protein (Takahashi and Ishii, 1952), and this material when polymerized caused much confusion before its identity was actually established, because it may simulate the virus itself, and can also take several other forms (Macleod *et al.*, 1963; Markham *et al.*, 1963) (Fig. 3). A most striking instance of the hazards involved in this kind of self-assembling system is shown in the recent study of the mutant strain of this virus, the PM2 strain, which was isolated (Siegel *et al.*, 1962) after treatment of the type strain of tobacco mosiac virus with nitrous acid in order to produce mutants. The PM2 mutant (Wittmann, 1965) has two amino acid differences from the type virus (No. 28 Thr → Ile, No. 95 Glu → Asp). These substitutions, though they may appear trivial, have the effect of modifying the shape of the protein slightly, so that, instead of forming a compact helix having $16\frac{2}{3}$ peptide molecules/turn which can enclose the nucleic acid in an effective manner, it polymerizes to form an extended narrow helix that has the wrong shape to cover the nucleic acid, which therefore is left bare (Siegel *et al.*, 1966) (Fig. 4). Naturally, this strain of virus is not only non-invasive, but is essentially a laboratory curiosity.

The turnip yellow mosaic virus, though the first to be recognized as a virus causing the production of incomplete empty particles, is by no means unique, the related wild cucumber mosaic virus being even less efficient in the use of its protein (Macleod and Markham, 1963). This lack of efficiency, however, is also shown by the turnip yellow mosaic virus when its host is treated with 2-thiouracil, a compound which interferes with nucleic acid synthesis but does not affect the production of coat protein (Francki and Matthews, 1962). Examination of the wild cucumber mosaic virus in crude sap specimens has revealed that even more bizarre objects are also produced. These are large spherical bodies, presumably made from virus protein which themselves enclose empty virus shells (Hitchborn and Hills, 1965) (Fig. 5). Recently, an even more remarkable set of structures has been found in plants infected with a necrotic isolate of turnip yellow mosaic virus (J. H. Hitchborn and G. J. Hills, unpublished). These structures look like thin flat plates under the electron microscope (Fig. 6) but examination of the negative by means of optical diffractometry (Fig. 7) shows fairly clearly than *in vivo* they are probably tubes. Purified suspensions of these tubes have been made, and it has been found that they have the same amino acid composition as that of the virus iself, and so presumably are made of virus protein. Very little definite detail can be seen on the tubes themselves, although the optical diffraction patterns (transforms) (Fig. 7) indicate that they have an exceptionally perfect structure. The reason for this is that the tubes are based on a helical arrangement of protein subunits, and since the tubes are flattened by the drying process, one has two patterns superimposed on each other out of register, so that each is obscured by the other.

It is actually possible to take advantage of the regularities in the structures of this kind to isolate the image from one side of the tube by a photographic technique (Markham *et al.*, 1964). This depends upon the fact that if one takes a photograph of a repeating structure, moves the structure on by exactly one repeat interval and photographs again, and so on, one can build up an image containing the repeating structure, while any other structure present and not having either the same repeat distance, or which is aligned in a different direction is automatically smoothed out. The procedure is carried out by semi-automatic apparatus, and is essentially equivalent to tuning into a station on the radio. When examined in this way (Fig. 8) the structure of one side of the tubes is revealed with unexpected clarity, and shows that these are made up from the peptide hexamers which form part of the virus shell (Fig. 9), which itself has twenty such hexamers and twelve pentamers, consisting in all of 180 peptides.

The above results show that the protein part of some small viruses, by virtue of their inherent self-polymerizing potentialities may form *in vivo* a number of structures other than virus particles, namely virus-size empty shells, large empty shells and long helical tubes. An even more extreme case is that of the cowpea chlorotic mottle virus, in which the unusual structures are not found *in vivo*, presumably because they are somewhat unstable under these circumstances (Bancroft *et al.*, 1967). This virus (Fig. 10) is one of a small number known to exhibit instability at neutral pH. Under these conditions salts, e.g. sodium chloride, at moderate strengths disrupt the virus reversibly into protein and nucleic acid. At neutrality, in the presence of nucleases, however, the nucleic acid is attacked, the virus falls to pieces, and is then reassembled into a large number of types of particle, all of which resemble viruses, and which include three kinds of spheres, one ellipsoidal particle and two differing kinds of tubes (Fig. 11). All of these contain some degraded nucleic acid, and, in the complete absence of nucleic acid, only one product is formed, a sphere resembling the virus, but consisting of two concentric layers of protein (R. Frist, unpublished). The general behaviour of the protein of this virus suggests that, in this particular instance, the formation of real virus particles depends absolutely on the meeting between the intact viral nucleic acid and the requisite 180 peptides, which are probably already organized into rings of six. The formation of the requisite number (12) of rings of five to form a particle, presumably follows by elimination of single peptides, and the structure so produced is fairly stable at the pH of cell sap, and therefore predominates.

This remarkable behaviour of the cowpea chlorotic mottle virus protein, does throw some light on the potentialities of self polymerizing proteins, and it also suggests why *in vivo* only the virus particle is to be found. The ability of the protein to form tubular structures is also interesting and suggests a possible origin for the similar tubes which have been found in a number of mammalian virus infections, such as, for example, that caused by human wart virus. The protein seems to be positively attracted to acidic surfaces primarily, of course, those of nucleic acids, but the surfaces of most viral proteins are themselves acidic so that concentric or coaxial shell formation takes place quite readily and, indeed, the cowpea chlorotic mottle virus protein will actually form shells around intact tobacco mosaic virus particles.

Apart from the structures already mentioned, there exists a large number of other bodies which have been found in virus-infected plants but in most of these the precise nature of these "inclusions" is not yet known. Many seem to be bags, presumably made of protein, which contain virus particles in large numbers (Fig. 12) (Markham, 1966). They may, in fact, be made from viral protein, which has polymerized spontaneously around groups of virus particles. Whatever they are, they certainly contribute greatly to the difficulty of isolating certain viruses.

REFERENCES

Aposhian, H. V. and Kornberg, A. (1962). *J. biol. Chem.* **237**, 519.
Bancroft, J. B., Hills, G. J. and Markham, R. (1967). *Virology* **31**, 354.
Bello, L. J., Van Bibber, M. J. and Bessman, M. J. (1961). *J. biol. Chem.* **236**, 1467.
Bovét, J. M. (1967). Thesis, Faculty of Sciences, Paris (C.N.R.S. No. A.O. 1289).
Bozarth, R. F. and Diener, T. O. (1963). *Virology* **21**, 188.
Diener, T. O. and Dekker, C. A. (1954). *Phytopathology* **44**, 643.
Francki, R. I. B. and Matthews, R. E. F. (1962). *Virology* **17**, 367.
Hitchborn, J. H. and Hills, G. J. (1965). *Virology* **26**, 756.

Johnson, M. W. and Hills, G. J. (1963). *Virology* **21**, 517.
Johnson, M. W. and Markham, R. (1962). *Virology* **17**, 276.
Kassanis, B. and Nixon, H. L. (1961). *J. gen. Microbiol.* **25**, 459.
Macleod, R., Hills, G. J. and Markham, R. (1963). *Nature, Lond.* **200**, 932.
Macleod, R. and Markham, R. (1963). *Virology* **19**, 190.
Markham, R. (1966). *Br. med. Bull.* **22**, 153.
Markham, R., Frey, S. and Hills, G. J. (1963). *Virology* **20**, 88.
Markham, R., Hitchborn, J. H., Hills, G. J. and Frey, S. (1964). *Virology* **22**, 342.
Markham, R. and Smith, K. M. (1949). *Parasitology* **39**, 330.
Mathews, C. K., Brown, F. and Cohen, S. S. (1964). *J. biol. Chem.* **239**, 2957.
Miczynski, K. A. (1959). *Acta biol. cracov.* **2**, 23.
Reichmann, M. E., Rees, M. W., Symons, R. H. and Markham, R. (1962). *Nature, Lond.* **195**, 999.
Sänger, H. L. and Knight, C. A. (1963). *Biochem. biophys. Res. Commun.* **13**, 455.
Siegel, A., Hills, G. J. and Markham, R. (1966). *J. molec. Biol.* **19**, 140.
Siegel, A., Zaitlin, M. and Sehgal, O. P. (1962). *Proc. natn. Acad. Sci. U.S.A.* **48,** 1845.
Smith, K. M. and Markham, R. (1945). *Nature, Lond.* **155**, 38.
Takahashi, W. N. and Ishii, M. (1952). *Phytopathology* **42**, 690.
Wildman, S., Cheo, C. C. and Bonner, J. (1949). *J. biol. Chem.* **180,** 985.
Wittmann, H. G. (1965). *Z. VererbLehre* **97.** 297.

Section 3c

ABBREVIATIONS USED IN THE FIGURES

A	aspartic acid + asparagine (protein hydrolysate)
γAB	γ-aminobutyric acid
ALA	alanine
ARG	arginine
ASP	aspartic acid
ASP(NH_2)	asparagine
CYS	cystine
FRUC	fructose
G	glutamic acid + glutamine (protein hydrolysate)
GLU	glutamic acid
GLUC	glucose
GLU(NH_2)	glutamine
GLY	glycine
GSH	glutathione
HIS	histidine
HOM	homoserine
ILE	isoleucine
LEU	leucine
LYS	lysine
MET	methionine
O-Ac HOM	*O*-acetyl homoserine
PAL	phenylalanine
PRO	proline
SER	serine
SUCR	sucrose
THR	threonine
TYR	tyrosine
VAL	valine
X	remaining amino compounds, (Fig. 16)

Physiological Aspects of Inorganic and Intermediate Nitrogen Metabolism (with special reference to the legume, *Pisum arvense* L.)

J. S. PATE

Botany Department,
Queen's University, Belfast, N. Ireland

INTRODUCTION

In practically every aspect of biology the research worker is faced with problems arising from the fact that the overall functioning of one particular organism may differ widely, sometimes basically, from that of other species. Study of the nitrogen metabolism of higher plants is no exception, and there is therefore much to be gained from an examination in depth of the physiology and biochemistry of certain selected species. From such studies it might be hoped eventually to distinguish the common and basic from the unusual features of nitrogen nutrition, and at the same time to determine the role played by specific nitrogen-containing compounds in the functioning of the individual plant and its constituent parts.

At Belfast, we have endeavoured to follow this line of approach in studies of the assimilation, transport and utilization of nitrogen in such species as the field pea (*Pisum arvense* L.), the cocklebur (*Xanthium pennsylvanicum* Wallr.) (Wallace, 1966; Wallace and Pate, 1967) and, recently, the hemiparasitic Angiosperm *Odontites verna* (Bell.) Dum. (Govier *et al.*, 1967). The present paper will be concerned primarily with the field pea, the most extensively studied of the above species.

THE NUTRITIONAL SYSTEM

The nutritional system displayed by the field pea (Fig. 1) may be regarded as being typical of that group of plants in which roots normally function as sites of assimilation of nitrate. In these plants organic solutes can be shown to be transported in both the xylem and the phloem, these conducting systems allowing a daily movement of nitrogenous and other compounds between organs of all ages and classes.

The fixation of carbon dioxide by photosynthesizing leaves of field pea may be considered as the primary component of its nutritional system.

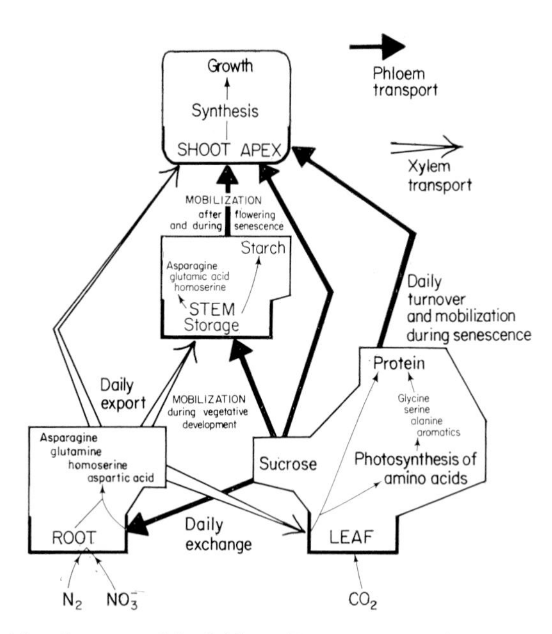

FIG. 1. The nutritional system of the field pea (*Pisum arvense* L.) as exhibited by nodulated plants grown in the absence of combined nitrogen or by uninoculated plants grown on a moderately low level of nitrate. (Adapted from Carr and Pate, S.E.B. Symposium, 21)

During photosynthesis of leaves a wide variety of soluble and insoluble substances is formed, including the major fraction of the amino acids and protein synthesized by the plant. If a leaf is fully expanded, it also provides a bulk source of carbon—sucrose—to heterotrophic organs of the plant (Fig. 1).

The root of field pea normally functions as an important contributor to the organic nutrition of the plant. An active system for reducing nitrate can be demonstrated in roots receiving nitrate (Wallace and Pate, 1965), while if nodulated, atmospheric nitrogen is assimilated through the symbiotic activity of nodule bacteria. In either circumstance, sucrose supplied by the shoot functions as the immediate source of carbon for manufacture of nitrogenous compounds in the root (Pate, 1962; Pate and Greig, 1964) The root may be shown to generate an excess of organic nitrogen under practically all conditions of growth and, during periods of assimilation, it releases nitrogen-rich molecules such as amides and ureides to the shoot in the transpiration stream (Pate and Wallace, 1964). Roots of the field pea export a fairly restricted range of amino acids, particularly those belonging to, or closely related to, the aspartate and glutamate families of compounds. In other species, synthesis in roots may show an even greater specialization with only one or two compounds carrying practically all of the nitrogen transported to the shoot (see Pate, 1962; Wallace and Pate, 1967).

A particularly interesting aspect of the metabolic system of the shoot concerns the processing of the amino acids and amides that it receives from the root. Each metabolite has a special significance in the economy of the shoot (Pate *et al.*, 1965). Certain compounds, for example asparagine, homoserine and glutamic acid, are mostly incorporated into soluble reserves and do not become generally available to growth until the tissues which store them have become senescent (Fig. 1). Other compounds, for example aspartic acid, glutamine and the leucines, are effective as immediate donors of carbon and nitrogen to synthesis of proteins in the leaf or in the growing parts of the plant. The place of amides in the daily functioning of the plant must be emphasized. In nodulated plants or in plants experiencing a moderately low supply of nitrate the amido and amino groups of amides must be regarded as the principal sources of nitrogen for *de novo* synthesis of amino acids. During photosynthesis in leaves, amides are consumed and amino acids such as alanine, glycine, serine and certain of the aromatic amino acids are produced (Pate, 1966). Indeed the synthetic capacities of root and leaf are complementary, roots providing amides and related amino compounds, leaves producing sucrose and amino acids not available in quantity from the root. However, nutritional interdependence of root and shoot is not obligatory, since, if large amounts of nitrate are supplied to the rooting medium, roots will progressively lose their capacity to synthesize amino compounds. In such a situation roots export nitrate to the shoot, and the latter must synthesize practically all of the amino acids required by the plant. This is indeed the situation that normally exists in species whose roots appear to be ineffective in assimilating nitrate, e.g. in *Xanthium pennsylvanicum* and *Perilla fruticosa* (Wallace, 1966; Wallace and Pate, 1967).

It is of obvious importance in any study of plant growth to examine the physiological processes which allow organic nutrients to be translocated to meristems and young organs of the shoot. In the field pea (Fig. 1) young organs can be shown to receive a varied nitrogenous diet derived directly or indirectly from every mature organ of the plant. These young organs are also themselves active in synthesis of certain amino compounds from the sucrose and amides that they receive from other parts of the plant (Pate 1966).

LEAVES AS ORGANS OF SYNTHESIS AND EXPORT

Throughout the life of the plant as successive leaves develop, each in turn makes specific demands from, or contributions to, the rest of the plant.

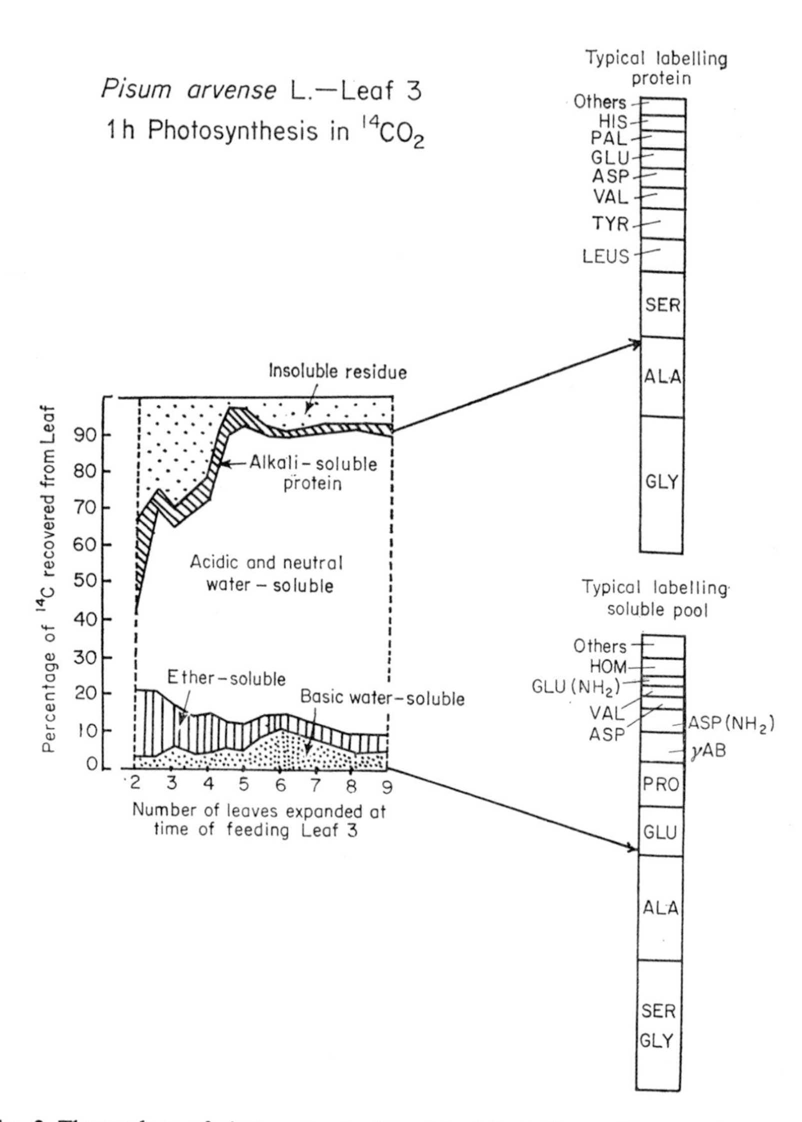

FIG. 2. The products of photosynthesis of Leaf 3 of the field pea. The age of the leaf at the time of feeding with $^{14}CO_2$ is given in terms of the number of fully expanded leaves on a shoot. Typical labelling patterns of free and protein-bound amino acids are given at the right of the figure. (Data from Brennan, 1966 and Pate, unpublished)

* Symbols for amino acids are listed on p. 218.

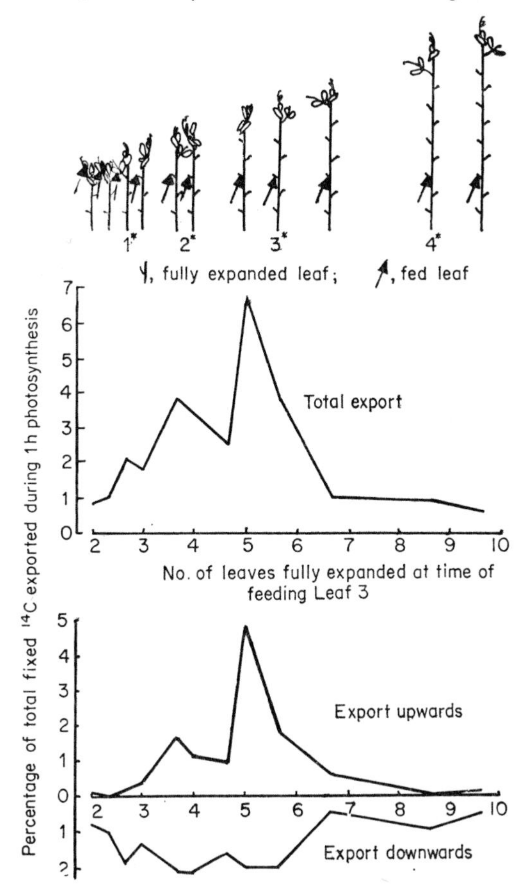

FIG. 3. The export of photosynthetically-fixed carbon from Leaf 3 of the field pea at different times during the life of the leaf. The age of the leaf at time of feeding is measured with reference to the number of fully expanded leaves on the shoot. Note the maxima in upward export during periods of expansion of Leaves 4 and 5. Also note that when the leaf is old it transports predominantly towards the roots (see also Fig. 5).

1*, 2*, 3*, 4* refer to the four stages depicted in Fig. 4. (Adapted from Carr and Pate, S.E.B. Symposium 21)

The synthetic capacities of the leaf change markedly during ageing. Young leaves at first have an absolute requirement for carbon, particularly carbohydrate. As a general rule, their photosynthesis results in carbon being fixed mainly into insoluble compounds and ether-soluble materials. Later, when fully expanded, the synthetic capacities of the leaf appear to be directed towards photosynthesis of sugars, particularly sucrose (Brennan,

1966) (Fig. 2). Free amino acids continue to be produced in quantity throughout the life of the leaf, and the distribution of radiocarbon among these always shows a bias towards amino acids typical of photosynthesis, e.g. glycine, serine and alanine (Fig. 2). Protein is labelled in photosynthesis largely through these same amino acids. The turnover of protein in photosynthesis is particularly high in young leaves, and it is significant that these same young leaves are also very active in nitrate reduction (Wallace and Pate, 1965).

Daily measurements of the export of photosynthetically-fixed radiocarbon from individual leaves offer a useful means of assessing their significance in the daily functioning of the plant. Certain well-defined changes occur in export activity during the life of the leaf. In early life the leaf is wholly or partially dependent on assimilates from older leaves, and it does not commence to export appreciable amounts of carbon until it has grown to a size equivalent to one quarter of its final area. Thereafter it exports assimilates in turn to the two leaves and two stem segments developing above it, exporting most to the younger of these organs (Fig. 3). Later in the life of the leaf, upward export virtually ceases: by this time, of course, other leaves higher up the stem have matured and are feeding the shoot apex. Assimilates move downwards at all times during the life of the leaf, a proportion of these moving laterally from conducting tissues to sites of storage in the stem, a proportion being consumed by the roots.

Sucrose is the main product to be exported from the leaf (Fig. 4). Labelled monosaccharides and labelled amino acids such as serine, glycine, leucine and asparagine can also be recovered from stem and petiole segments adjacent to a leaf which has been fed with $^{14}CO_2$, but it is impossible to determine whether these compounds have been transported from the photosynthesizing leaf or whether they have arisen secondarily by metabolism of labelled sucrose in the conducting elements and in adjacent tissues (Fig. 4). Moreover, in feeding experiments of long duration, stem tissues may become labelled with radiocarbon which has passed through the root system.

It is of some value to construct a picture of the combined export activities of all leaves of a plant and to show what changes occur as the plant develops from the vegetative to the reproductive condition. In young plants a free circulation of solutes occurs with upper and lower leaves supplying photosynthate to both roots and shoots (3- and 5-leaf plants, Fig. 5). Later, as more leaves develop (7- and 12-leaf plants, Fig. 5) nutrition becomes stratified in such a manner that upper leaves are effective in supplying assimilates to young leaves and to the shoot apex, lower leaves in supplying the roots, while leaves in the middle of the stem supply adjacent stem tissues and, to a lesser extent, the root and shoot apex (Brennan, 1966). These

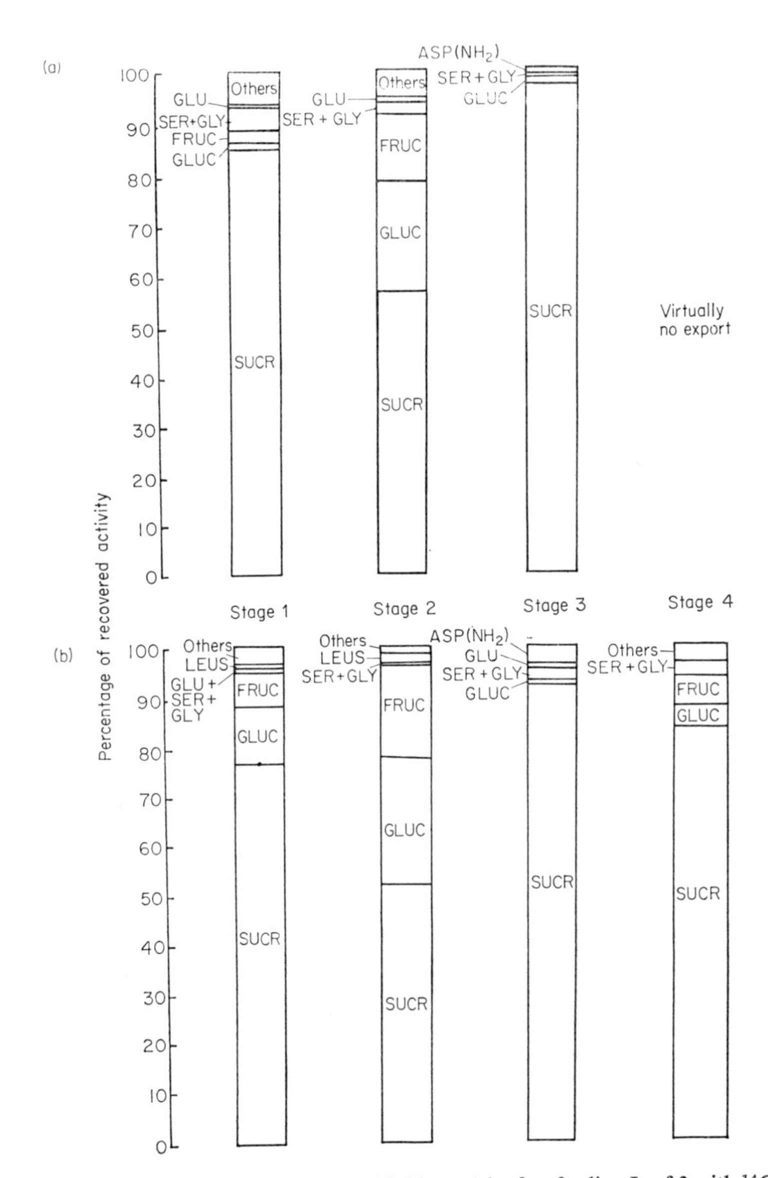

FIG. 4. ^{14}C-labelled compounds in stems of field pea 1 h after feeding Leaf 3 with $^{14}CO_2$. (a) an internode segment cut from immediately above the node subtending the fed leaf; (b) an internode segment cut from immediately below the node subtending the fed leaf. Stages 1–4 refer to specific times during the life of Leaf 3 (see Fig. 3). (Data from Brennan, 1966)

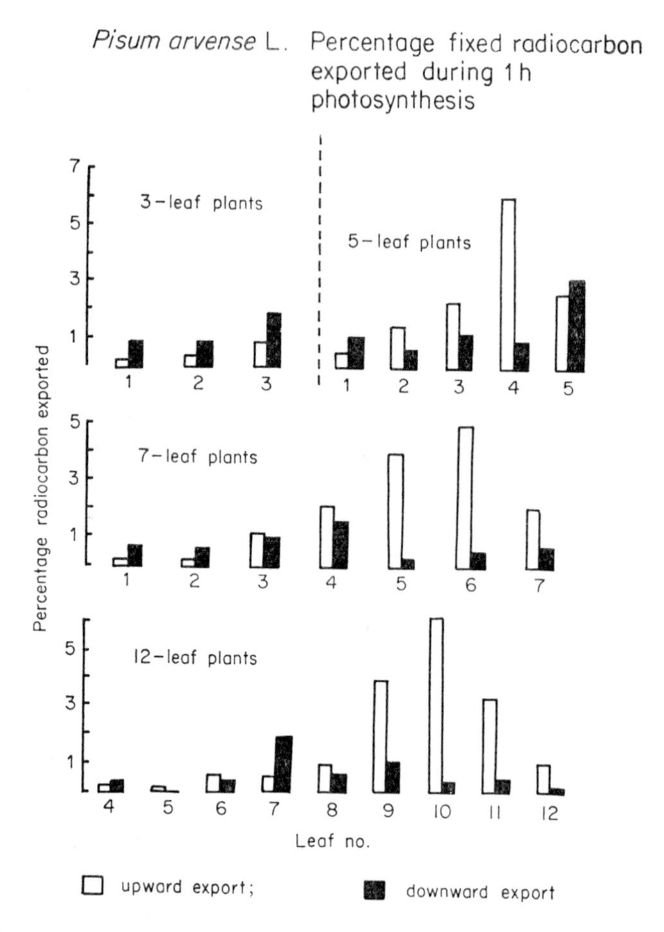

FIG. 5. Export of photosynthetically fixed carbon from leaves at four times during the vegetative life of the field pea. The data are derived from individual measurements of export from single leaves. (From Carr and Pate, S.E.B. Symposium, 21)

nutritional zones become more clearly delimited in flowering pea plants, where each leaf subtending a flower or flowers can be shown to export assimilates almost entirely to the flowers of its node (A. Flinn and J. S. Pate, unpublished). At this time, non-reproductive parts of the plant must be sustained by assimilates donated by leaves which do not subtend flowers. Once these latter leaves senesce, death of whole sections of the root system may be observed.

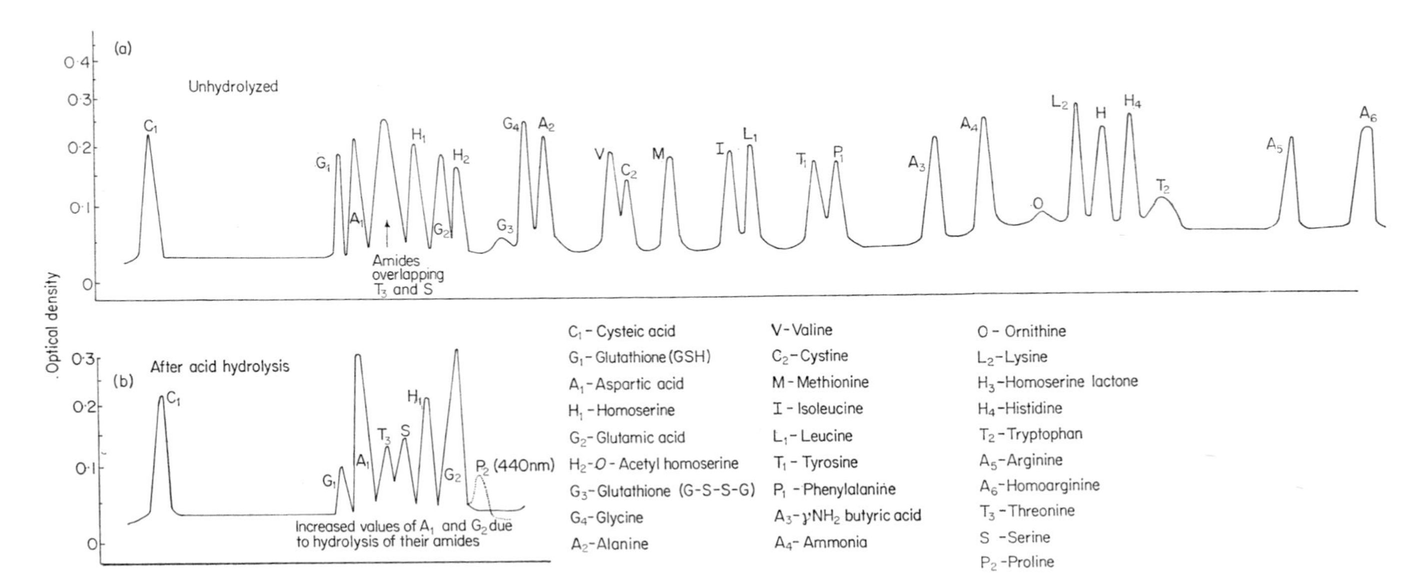

FIG. 6. Ninhydrin positive substances separated by a $6\frac{1}{2}$ h "fast" chromatographic system developed for use with a Technicon amino acid analyzer. (a) Separation of an unhydrolyzed mixture of amino acids (each 0·1 μmole). (b) Part of a trace of the same mixture after 3 h hydrolysis in N HCl. The amides glutamine and asparagine have been converted to their parent acids and the increase in these acids over the amounts present in the unhydrolyzed extract is used as a measure of the amides. Threonine and serine are also measured from the hydrolyzed extract. (See Pate and Wallace, 1966)

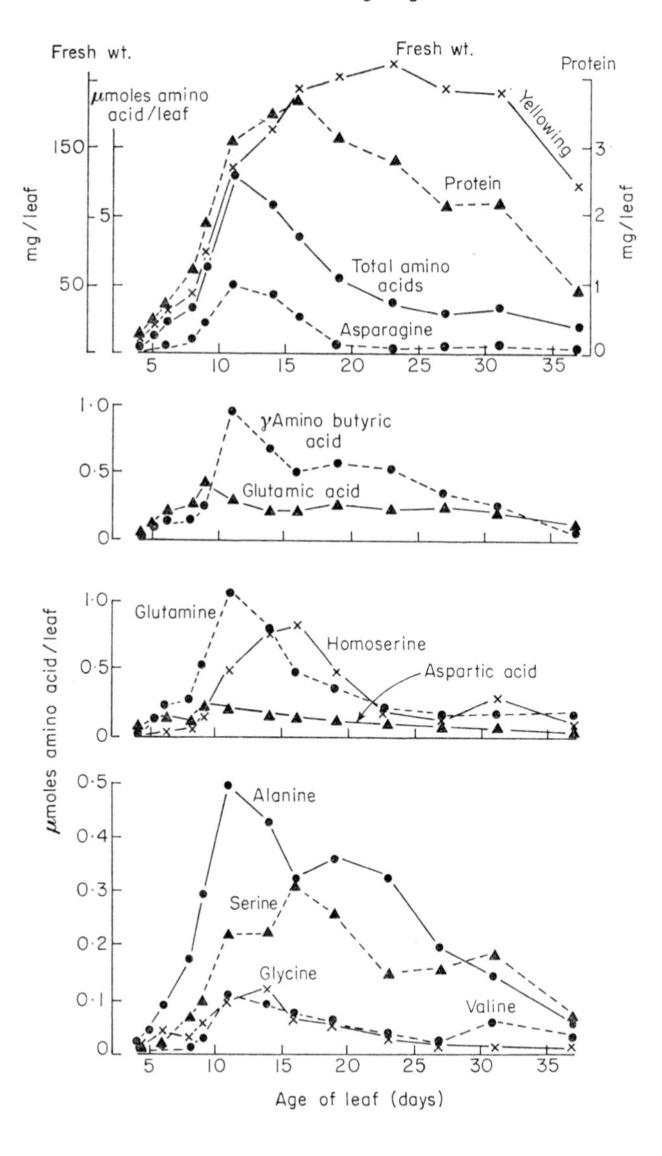

FIG. 7. Changes with time in the composition of Leaf 5 of the field pea. Each measurement was made from a sample of 100 leaflets, the experimental material being identical with that used for the study of Internode 5 described in Fig. 8. (Data from Carr and Pate, S.E.B. Symposium, 21)

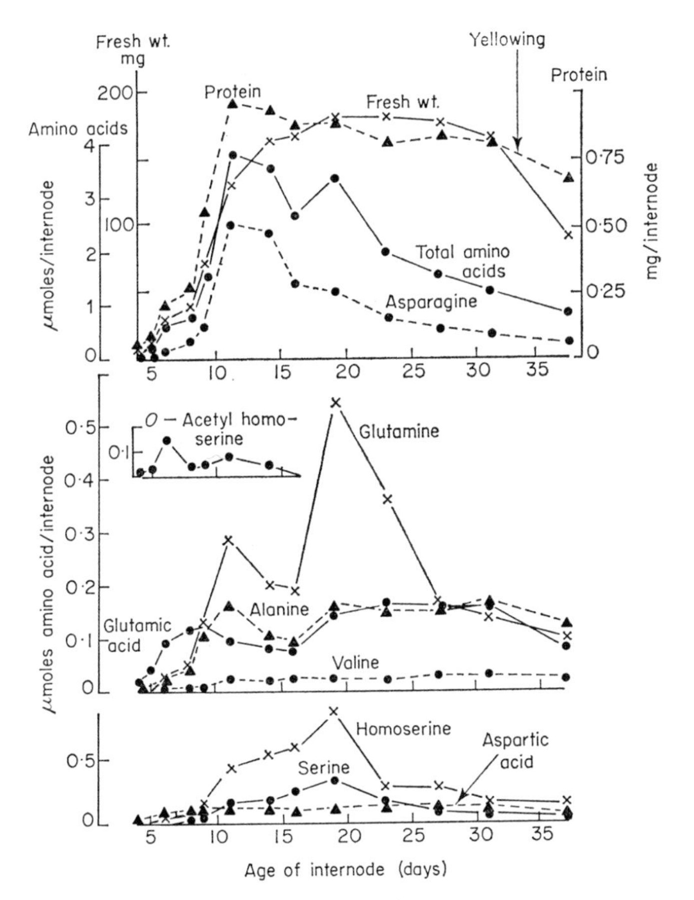

FIG. 8. Changes with time in the composition of Internode 5 of the field pea. Each measurement was made from a sample of 100 internodes, the experimental material being identical with that used for the study of Leaf 5 (Fig. 7). (Data from Carr and Pate, S.E.B. Symposium, 21)

LEAVES AND STEMS AS RESERVOIRS OF NUTRIENTS

In the symbiotic plant, nitrogen arriving from the root is stored entirely in organic form as protein or soluble amino compounds. A similar situation exists in uninoculated plants receiving an equivalent amount of nitrogen as nitrate, although such plants may accumulate free nitrate in their roots, their stem parenchyma and, rarely, in their leaves.

Each internode and leaf of the plant displays a highly characteristic pattern of changes in its soluble and insoluble constituents (see Carr and

Pate, 1967). The history of two organs, an internode and a leaf, will serve as typical examples from nodulated plants of field pea (Figs. 7 and 8). In both of these organs protein and amino acids continue to accumulate until the fresh weight is about three-quarters of its final maximum. There is then a gradual loss of soluble and insoluble nitrogen until senescence. A study of changes in individual amides and amino acids, using an amino acid analyzer (Figs. 6–8), has shown that each organ of the plant possesses a characteristic set of nitrogenous solutes, and that during its ageing a sequence of maxima is manifest in the amounts of individual amino compounds. The full significance of the changes that occur in the soluble fraction of the tissue has yet to be appreciated, but since the sequence attained under any particular nutritional regime is highly reproducible, amino acid analysis may be found to be useful in assessing the physiological age and current nutritional status of the plant.

It is of special interest to note that all vegetative organs commence to show a net loss of nitrogen long before they are fully grown. These losses, and those occurring later in turnover and mobilization of nitrogen, must be regarded as important elements in the nutrition of meristems and young parts of the plant. It is evident that at any one time during growth mature organs of different physiological age and nutritional balance will be contributing specific commodities to younger regions of the shoot. Extremely little is known of the mechanisms and the species of compounds involved in the mobilization of the nitrogenous reserves of mature organs of the plant; even less is known about the way in which the overall senescence of these organs is integrated with the vital functioning of younger parts of the plant body.

ROOTS AS ORGANS OF SYNTHESIS AND EXPORT

Seedlings of field pea grown on a constant supply of nitrate exhibit high levels of nitrate reductase. The bulk of this enzyme is present in young sub-apical regions of primary and lateral roots. It is difficult to detect the enzyme in older parts of the root system (Fig. 9). The distribution of nitrogenous solutes along the length of lateral roots has been studied in seedlings receiving nitrate and in seedlings grown without nitrate. The results are presented in Fig. 9. Both types of seedling contain highest levels of soluble nitrogen in the youngest parts of their roots. There is, however, remarkably little difference between the amino acid content of roots given the two treatments, although the plants receiving nitrate were found to be exporting large amounts of organic nitrogen to their shoots. It may therefore be concluded that roots are not programmed to store large quantities of soluble nitrogen and that the products of their assimilation are immediately released to the shoot.

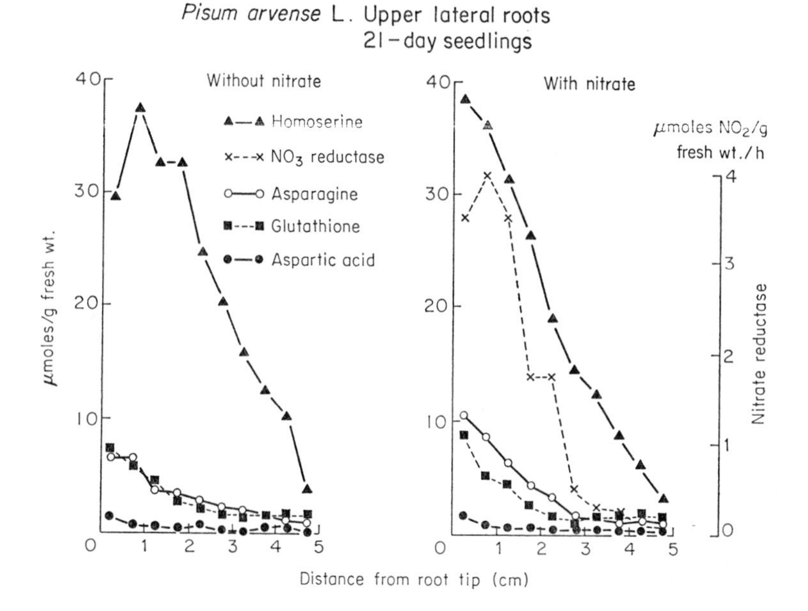

FIG. 9. The distribution of the major constituents of the pool of soluble nitrogen in roots of field pea grown in the presence or absence of nitrate. Data for nitrate reductase are included from plants grown with nitrate. (From Carr and Pate, S.E.B. Symposium, 21)

Induction of nitrate assimilating activity in roots has been followed after presenting nitrate to seedlings of field pea (Wallace and Pate, 1965). Within two hours of applying nitrate to the rooting medium, nitrate reductase can be detected in both the root and shoot, and during the next 24 hours the rate of export of organic nitrogen from the root increases from the very low level typical of nitrogen deficiency to a level equal to that of effectively nodulated plants of comparable age (Fig. 10). Throughout this period little increase occurs in the amount of soluble organic nitrogen in the roots, but substantial increases may be observed in the amount of organic nitrogen in the shoot (W. Wallace, unpublished).

Nitrate absorbed in excess of the assimilatory capacities of the root is either stored as such in the root or escapes to the shoot *via* the transpiration stream (Wallace and Pate, 1965). It has been shown that where low levels of nitrate are maintained in the rooting medium (e.g. 10 ppm NO_3^-) the root will continue to function as the major site of reduction, and its ability to assimilate nitrate will become strengthened as the seedling grows (Fig. 11). On the other hand, in the presence of higher levels of nitrate (50–500 ppm NO_3^-) the assimilatory capacities of the root will decline

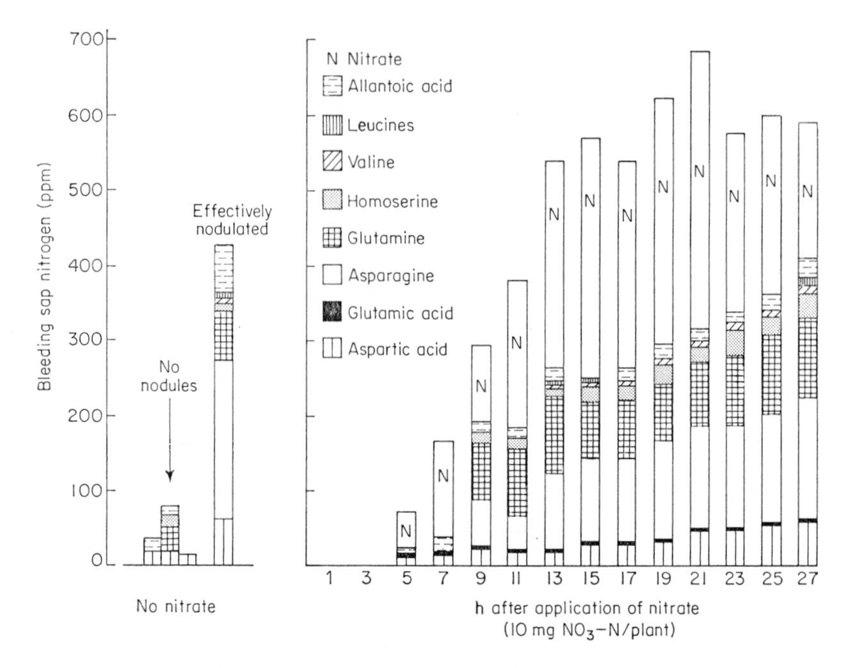

FIG. 10. Progressive changes in the composition of bleeding sap following application of nitrate to the roots of uninoculated, four-leaf seedlings of field pea. Sap composition is also recorded for untreated plants and for effectively nodulated seedlings of similar age. (From Wallace and Pate, *Ann. Bot.* **29**, 655)

with time, the shoot eventually assuming the dominant role in nitrate reduction. In the latter situation roots may be shown to become progressively starved of carbohydrate.

Although it has been shown that roots do not function in bulk storage of amino nitrogen, it is still of interest to know whether the compounds produced by root cells during assimilation of inorganic nitrogen are all equally available for export to the shoot. Comparison of the distribution of nitrogen in the pool of soluble nitrogen of whole roots with that in sap bleeding from identical roots would suggest that export is a selective process (Fig. 12). Certain compounds such as glycine, alanine, glutamic acid and γ-amino butyric acid appear to be stored preferentially by the root, others such as homoserine, the amides and aspartic acid are more readily available for release to the transpiration stream. Labelling of roots with $^{35}SO_4^{2-}$ reveals a similar picture, ^{35}S-methionine being freely released from the root, while ^{35}S-glutathione is accumulated in roots but is not readily exported (Pate, 1965).

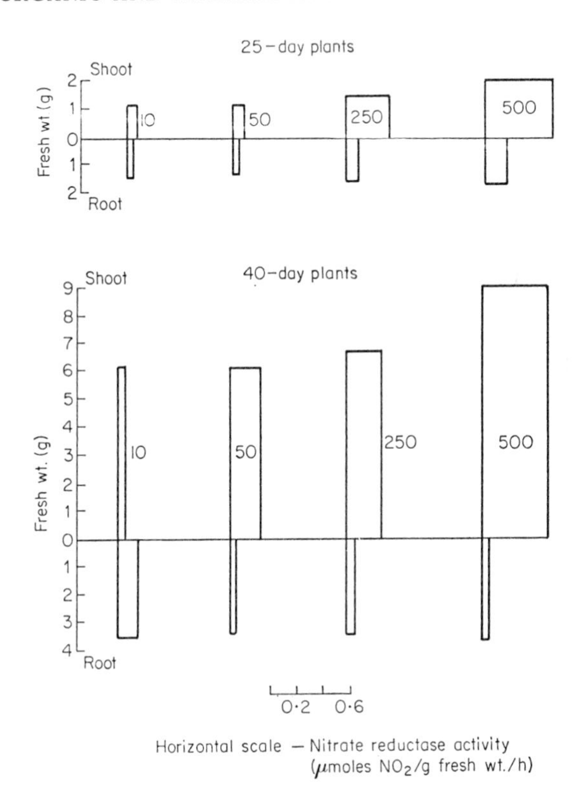

FIG. 11. Influence of constantly maintained levels of nitrate in the medium (10, 50, 250 or 500 ppm NO_3^-) on nitrate reductase activity in the root and shoot of the field pea. The area of each rectangle (fresh weight × specific enzyme activity) is proportional to the total reductase assayed from a particular root or shoot. (Data from Wallace and Pate, 1967)

The cycling of carbon *via* the roots is an obligatory feature of the physiology of the nodulated plant, and judging from rates of accumulation of nitrogen in shoots, it is a process which continues steadily throughout the life of the plant. The compounds moving from roots have been described for a series of stages in the life of nodulated plants (Fig. 13), and it has been shown that the two amides glutamine and asparagine may carry 80–90% of the nitrogen assimilated during the plant life cycle. Synthesis in roots requires a considerable expenditure of carbohydrate, and once plants have developed ten or more leaves it has been estimated that the daily export of carbon from a root to its shoot in the form of amino acids and amides may sometimes exceed the loss of carbon by respiration (J. S. Pate, unpublished).

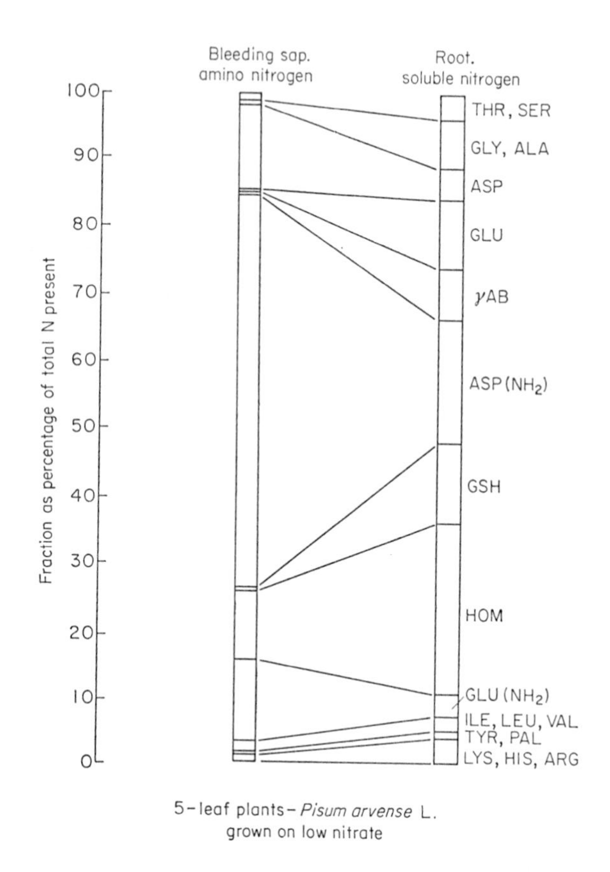

FIG. 12. A comparison of the composition of the nitrogenous fractions of bleeding sap and the ethanol-soluble fraction of roots of field pea. (Wallace and Pate, unpublished)

THE NUTRITION OF THE SHOOT APEX

It is apparent from the foregoing discussion that the sustenance of shoot meristems and juvenile organs of the shoot represents a complex series of processes involving the concerted activity of all mature parts of the root and shoot and biochemical reactions which would appear to be intimately linked with the primary synthesis of the root.

One event of this nutritional sequence which is amenable to study is the initial processing of compounds arriving from the root. The fate of the carbon from these compounds has been recorded after single ^{14}C-labelled amino compounds have been fed to detached shoots *via* the transpiration stream (Brennan *et al.*, 1964; Pate *et al.*, 1965). The main finding from such

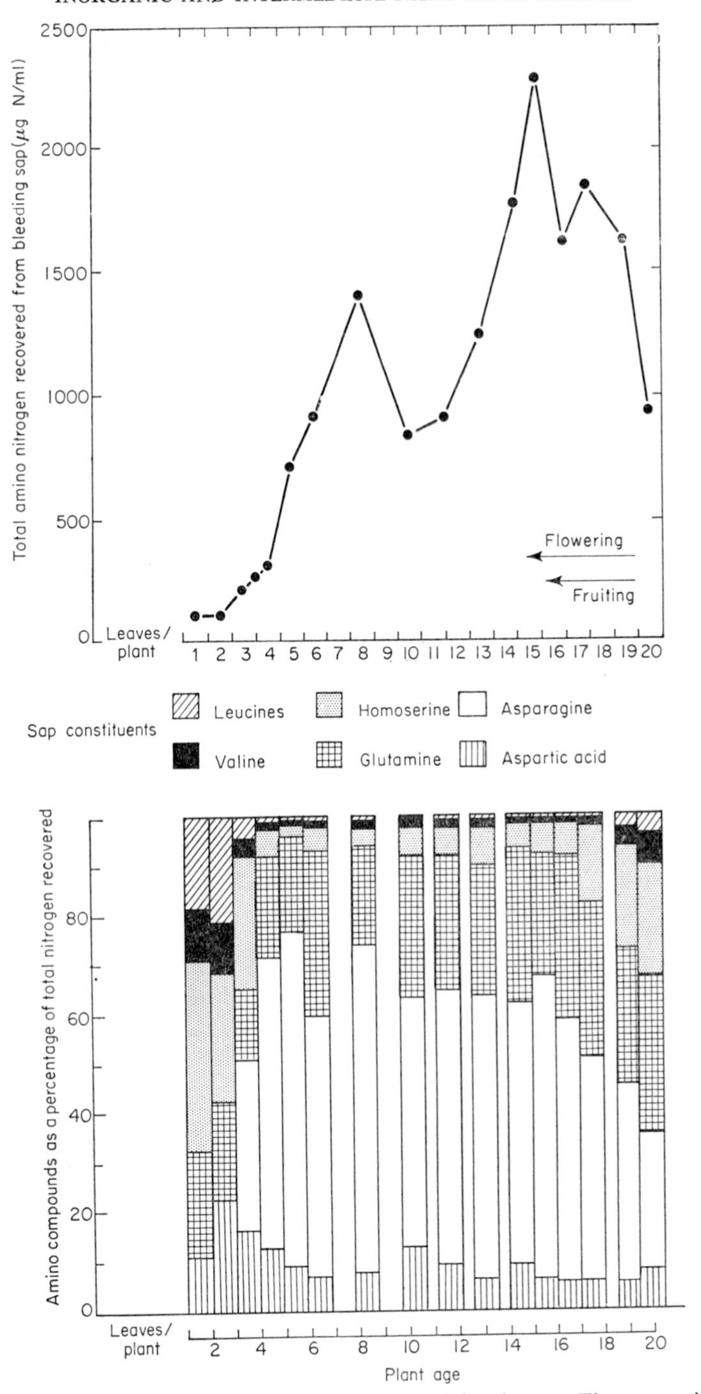

FIG. 13. Export of nitrogenous compounds from nodulated roots. The composition of the nitrogenous fraction of bleeding sap of field pea collected at different times in the life of nodulated plants raised in nitrogen-free medium. 40–50 plants were decapitated at each time of sampling and bleeding sap was collected from their roots for a 2h period centred on noon. (From Pate *et al.*, *Ann. Bot.* **29**, 475)

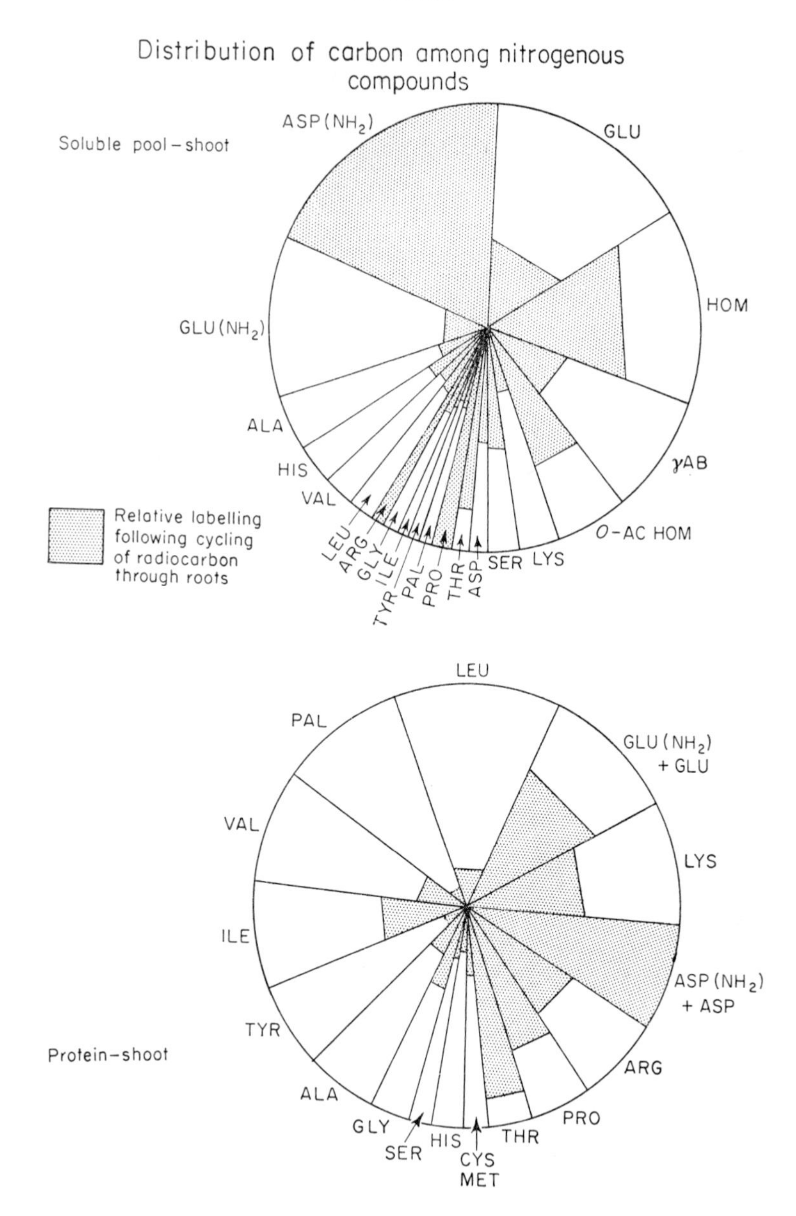

FIG. 14. The significance of the root in donating carbon to amino compounds of the protein and soluble phase of the shoot of field pea. Sectors of each circle are arranged so that their areas are equivalent to the relative amounts of carbon present in the individual amides and amino acids. The degree to which the sector representing an individual compound is stippled is in direct proportion to the ease with which that compound receives carbon which has cycled through the root system.

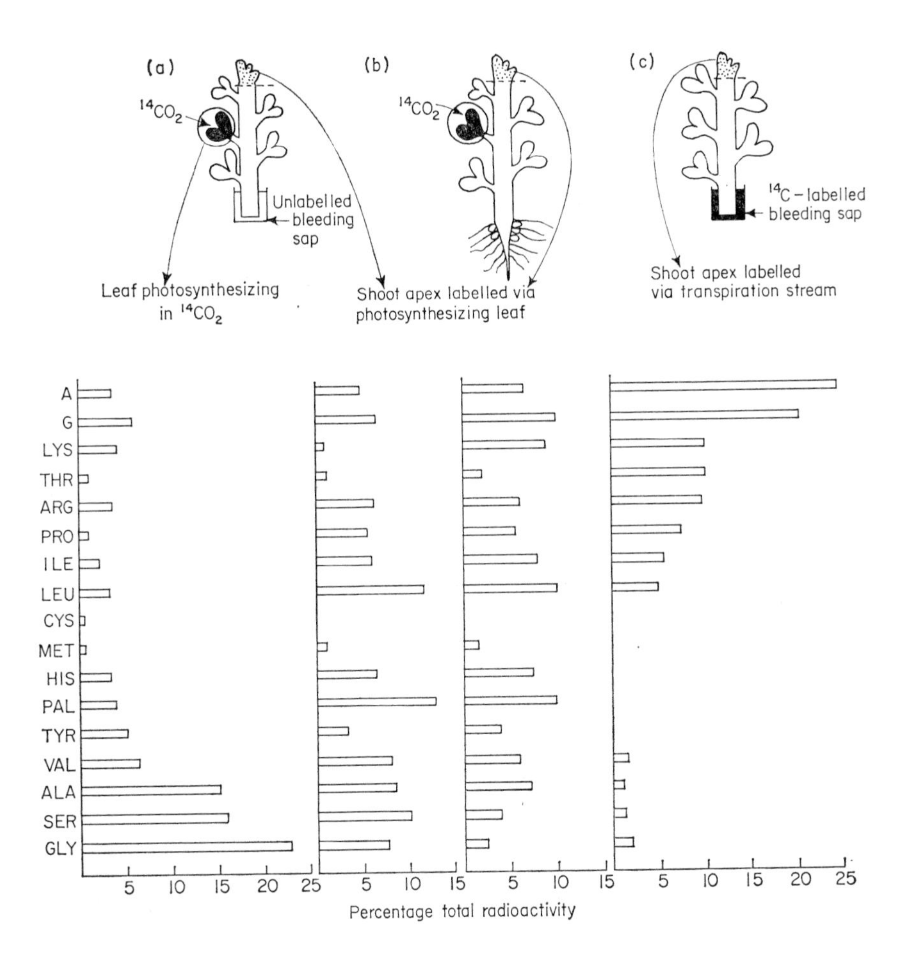

FIG. 15. The nutrition of the shoot apex. I. Utilization of carbon derived from root and leaf in synthesis of proteins at the apex. Percentage distribution of radiocarbon among the amino-acids of alkali-soluble protein from Leaf 3 or the shoot apex of five-leaf plants of field pea. (a) Leaf 3 of each shoot was illuminated in the presence of $^{14}CO_2$ while the cut end of the shoot absorbed diluted, unlabelled bleeding sap. (b) Fed with $^{14}CO_2$ as in (a) but using complete, nodulated plants. (c) Shoots retained in an unlabelled atmosphere while ^{14}C-labelled sap was fed *via* the transpiration stream. Data obtained for plants harvested after 48 h. (From Pate, *Ann. Bot.* **30**, 93)

experiments is that the bulk of the nitrogenous materials arriving from roots first enters lower leaves or is abstracted by older internodes and petioles. Uptake by these mature parts of the shoot is found to be greatly in excess of their current requirements, and much of the absorbed material is immediately processed and eventually re-exported to growing organs at the top of a shoot. Transfer from older to younger parts of the shoot presumably takes place in the phloem. It has been shown that amino acids, released from turnover of protein, move from these old organs as well as nitrogenous materials mobilized from soluble reserves (Pate *et al.*, 1965).

The significance of the root as a nutritive organ is shown diagrammatically in Fig. 14. In this figure the amino acids of the soluble pool and protein of the shoot are represented as sectors whose areas are proportional to relative contents of carbon. Areas of stippling in the sectors of each circle are displayed, these stippled areas being proportional to the relative ease with which the various compounds accept carbon which has passed through the root. It can be seen that root metabolism makes a relatively large contribution to certain amino compounds of the soluble pool of the shoot—notably asparagine, homoserine, arginine, proline and threonine; and to certain amino acids of protein—particularly aspartyl and glutamyl residues, proline, lysine and arginine. Transfer of carbon from the root to other amino compounds in the shoot takes place much less readily and other processes in the shoot, for example, photosynthesis, or metabolism of carbohydrate or protein, must be held reponsible for synthesis of the bulk of these compounds. A proportion of the ^{14}C-labelled materials generated by a root becomes incorporated into protein of the shoot apex. Distribution of labelled carbon among protein amino acids from this region is remarkably similar to that of the body of the shoot (cf. Fig. 15, right-hand column, and Fig. 14, shoot protein).

A second element of the nutrition of the shoot which has been studied is the direct transfer of photosynthetically-fixed materials from leaves to the shoot apex. This is best investigated by feeding upper leaves with $^{14}CO_2$, since these leaves are most likely to be supplying sucrose and possibly small amounts of amino acids directly to the shoot apex (Figs. 4 and 5). When plants are fed in this manner the proteins of the shoot apex become generally labelled among their amino acids, labelling of each amino acid being fairly well in proportion to its relative abundance in the protein (Fig. 15, left and middle columns). Thus the apex itself appears to metabolize sucrose actively and, using a nitrogen source such as an amide, is capable of synthesizing all of the amino acids required for protein.

Photosynthesis by lower leaves on a shoot also makes a measurable contribution of carbon to protein of the shoot apex. However, the nutritional role of these leaves is indirect. They export sucrose mainly to the roots, and after cycling of some of this carbon has occurred through the roots, protein amino acids typical of root metabolism become incorporated into proteins forming at the shoot apex. Indeed, in the intact plant it has been shown that each leaf on a shoot makes a measurable contribution of

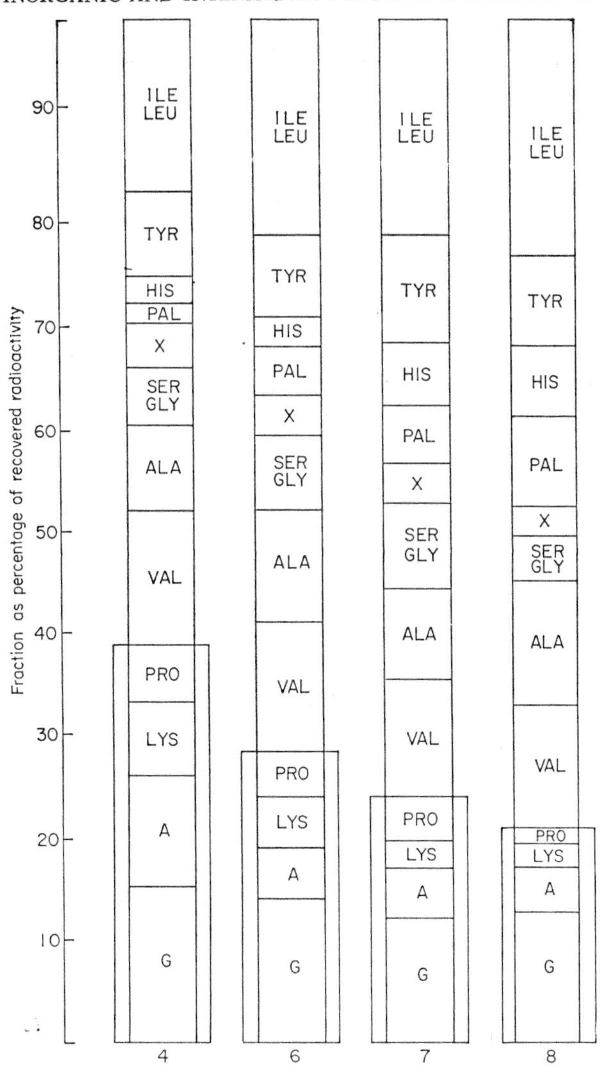

FIG. 16. The nutrition of the shoot apex. II. The contribution of photosynthetically-fixed carbon from individual leaves to protein of the shoot apex of the field pea. Percentage distribution of radiocarbon among the amino-acids of alkali-soluble protein from the apical portion of the shoot of nodulated, 8-leaf plants of field pea, following feeding of single leaves of different age with $^{14}CO_2$. Leaves are numbered from base to apex of the shoot, Leaf 4 being the oldest healthy leaf, Leaf 8 the youngest fully expanded leaf. Six plants were used in each treatment, each being fed with 100 μc $^{14}CO_2$. Plants were harvested 48 h after application of the isotope. Note that lower leaves contribute relatively more carbon to amino acids typical of synthesis by roots (see "bracketed" amino compounds). Much of the carbon from these lower leaves is cycled through the root system. (From Pate, *Ann. Bot.* **30**, 93)

photosynthetically-fixed carbon to the shoot apex, lower leaves contributing carbon mainly to the protein amino acids glutamate, aspartate, proline and lysine; middle and upper leaves contributing less to these compounds but relatively more to the compounds derived from materials arriving via a direct phloem pathway to the apex (Pate, 1966; Fig. 16). It is gratifying to find in these results for the complete plant a full measure of agreement with earlier investigations on the metabolism of isolated shoots, leaves and roots.

ACKNOWLEDGEMENT

This work has been supported by a grant from the Agricultural Research Council.

REFERENCES

Brennan, H., (1966). M.Sc. Thesis, Queen's University, Belfast.
Brennan, H., Pate, J. S. and Wallace, W. (1964). *Ann. Bot.* **28**, 527.
Carr, D. J. and Pate, J. S. (1967). *In* "Aspects of the biology of ageing", S.E.B. Symposium, 21, p. 459, Cambridge University Press.
Govier, R. N., Nelson, M. D. and Pate, J. S. (1967). *New Phytol.* **66**, 285.
Pate, J. S. (1962). *Pl. Soil* **17**, 333.
Pate, J. S. (1965). *Science, N.Y.* **149**, 547.
Pate, J. S. (1966). *Ann. Bot.*, **30**, 93.
Pate, J. S. and Greig, J. M. (1964). *Pl. Soil* **21**, 163.
Pate, J. S., Walker, J. and Wallace, W. (1965). *Ann. Bot.* **29**, 475.
Pate, J. S. and Wallace, W. (1964). *Ann. Bot.* **28**, 83.
Pate, J. S. and Wallace, W. (1966) *In* "4th Amino Acid Colloquium", Technicon Instrument Company Ltd.
Wallace, W. (1966). Ph.D. thesis, Queen's University, Belfast.
Wallace, W. and Pate, J. S. (1965). *Ann. Bot.* **29**, 655.
Wallace, W. and Pate, J. S. (1967). *Ann. Bot.* **31**, 213.

Section 3d

The Effect of Climate and Time of Application of Fertilizers on the Development and Crop Performance of Fruit Trees

D. G. HILL-COTTINGHAM

Long Ashton Research Station
University of Bristol, England

INTRODUCTION

A mature cropping apple tree is a very complex organism with a large number of different growing points, the activity of which may vary widely with the season. For example the temperature, moisture, degree of aeration and pH of the soil can interact to affect root activity and efficiency; similarly air temperature and light intensity may alter the metabolism of the added nitrogen in leaves and flowers. Also it must be remembered that the performance of a fruit plant is judged not by its total growth but by the weight and quality of the final crop, and even a heavy crop is not always desirable, for it often means a reduction in fruit size and can lead to a tendency for biennial bearing. Hence, it is not surprising that the response of a particular tree to a given form of fertilizer nitrogen can vary so widely with the circumstances and time of year at which it is applied. The results of the large number of experiments carried out on one or more aspects of this complex problem have been reviewed recently by Boynton and Oberly (1966).

The conventional method of applying nitrogen fertilizer to apple trees is by a ground application in late winter but, recently, interest has been stimulated in the possible advantantages of applying it later in the season. The results reported below are based on data obtained at Long Ashton in such an experiment using maiden apple trees growing in soil to which fertilizer nitrogen was added in the spring, summer or autumn (Hill-Cottingham and Williams, 1967), and also on some hitherto unpublished analytical data. The results are discussed in relation to other recent work and literature.

VEGETATIVE GROWTH, FLOWERING AND FRUITING

The critical factor that determined most of the subsequent behaviour of the young apple trees (rootstocks plus bud) used in the experiments was

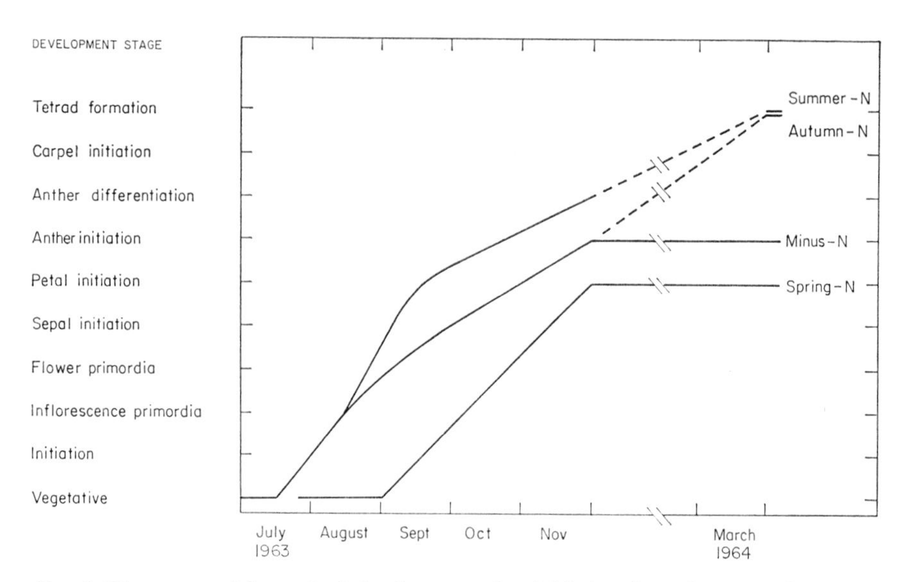

FIG. 1. The course of flower bud development after initiation from the vegetative stage.

the date of cessation of primary extension shoot growth. The duration of primary extension growth was found to be in proportion to the nitrogen status of the tree at the beginning of the season, as well as being dependent on the availability of fertilizer nitrogen during that period (Oland, 1959; Hill-Cottingham, 1963). The growth of the scion dominated the whole plant during this period, for no increases in total weight of stock or roots took place until it ceased and, moreover, all axillary buds remained solely vegetative. It was also found that as extension growth ceased, initiation and differentiation of flower primordia commenced, a correlation that had been noted previously (Williams, 1963). On the trees given no fertilizer nitrogen, flower initiation commenced in early August and differentiation proceeded steadily through the rest of that season, while on those given nitrogen in the spring it was delayed until the latter half of September, and might well have been later if the fertilizer had not caused root damage. Consequently, the flower buds on the trees given "spring-N" were less well advanced at the beginning of the winter than on the "minus-N" trees (Fig. 1). In the same experiment other trees that had been previously untreated were given fertilizer nitrogen after extension growth had stopped, either in mid-August, "summer-N", or in late October, "autumn-N". Significantly, these applications did not cause a renewal of extension growth, although girth increases and root growth were recorded, but they did have a profound effect on flower bud development which was most noticeable

TABLE I

Blossom bud production and fruit set records. Mean values/tree.

	Treatments			
	Minus-N	Spring-N	Summer-N	Autumn-N
Total number of nodes	29	39	29	29
Blossom buds initiated	12·9	15·5	14·2	13·7
Blossom buds that flowered	10·0	11·1	12·5	11·3
Total number of flowers	45·7	50·3	77·2	59·7
Final fruit set	0·5	0·1	2·7	14·5

the following spring. Then both the "summer-N" and "autumn-N" trees flowered together at least five days before the others. But the appearances of the flowers on the two treatments were very different, the large flowers and the large green primary leaves on the "summer-N" trees being most striking. From Fig. 1 it will be seen that, when the stimulus from the summer fertilizer application reached the buds, sepals and petals were being formed and this has resulted, apparently, in the differentiation of these tissues being accelerated. The "autumn-N" flowers were much smaller, in fact very similar in appearance to the "minus-N", despite the earlier development of the former, but the fruit-set results (Table I) indicated an important difference between them that is attributed to ovule longevity. It must be significant that the only organs to develop in the flower after the autumn fertilizer application are the ovules. In particular, the ovules of the "autumn-N" flowers were distinguishable by an extended period of meristematic activity in the nucellus after anthesis. It is known that meristematic tissue produces auxin (Leopold, 1964) and it is suggested that the enhanced production of auxin by the ovules themselves delayed their own senescence and death. In England climatic conditions during pollination are often marginal and accentuate the importance of the time necessary for pollen tube growth. Any method that increases ovule longevity thus leads to a disproportionate increase in the chances of good fruit set (Williams, 1965).

During blossoming there was a marked interval on the "autumn-N" trees between the appearance of the first leaves, which were obviously nitrogen deficient, and the darkening of these leaves, presumably due to the translocation of the abundant reserve nitrogen from the roots. These results suggest that the early resumption of activity in buds after the dormant period, which resulted in the significant morphological differences in the ovules referred to above, is not directly connected with nitrogen metabolism but indicates a stimulus from the roots reaching the flower buds in advance of the bulk of the nitrogen reserves. One explanation would be to postulate a substance, originating in roots stimulated by increased nitrogen availability, and capable of initiating or prolonging active cell division in potential flower buds or developing flowers. A cytokinin could fill this role and the presence of such substances has been indicated in the roots of grape-vines (Loeffler and van Overbeek, 1964), and application of synthetic cytokinins to dormant apple buds during the summer has been shown to cause renewed growth (Chvojka *et al.*, 1961).

THE TOTAL NITROGEN CONTENT OF FRUIT PLANTS

Many workers have investigated the seasonal changes in nitrogen content that take place in apple trees, but most have studied only the percentage composition of leaves or terminal shoots in isolation from the rest of the tree. More recently whole trees have been sampled to follow changes in distribution of their total nitrogen content (Mason and Whitfield, 1960; Poulsen and Jensen, 1964). However, little work has been carried out with whole trees in which the consequences of additional fertilizer nitrogen have been investigated. Exceptions to this include the reports of Oland (1959), Mori and Yamazaki (1959) and Delap (1967), all of whom used trees in sand irrigated with a nutrient solution, the results from which may not be comparable with trees in soil given a single fertilizer application.

In this experiment, at regular intervals from budbreak until blossoming the following year, whole trees were sampled and divided into morphologically different parts, which were then analyzed separately for total nitrogen. The seasonal changes in the overall amount of nitrogen in these trees, Fig. 2, show that in England fertilizer nitrogen can be absorbed by apple trees whether applied in spring, summer or autumn, but this overall picture hides the fact that the subsequent distribution of this nitrogen is not always the same. Nitrogen absorbed in spring or summer is translocated rapidly to all parts of the plant, but that absorbed after leaf-fall remains largely in the roots and stock throughout the winter.

In this same experiment it was noted that about 30% of the total nitrogen in all the trees was concentrated in the leaves at the end of the summer, in spite of the large differences in actual amounts between treatments. Mason and Whitfield (1960) have reported that approximately 35% of the total nitrogen in 6-year-old trees was present in the leaves, while the figure given for 2–3 year-old rootstocks by Poulsen and Jensen (1964) was even higher, about 50%. The migration of nutrients from leaves during senescence has

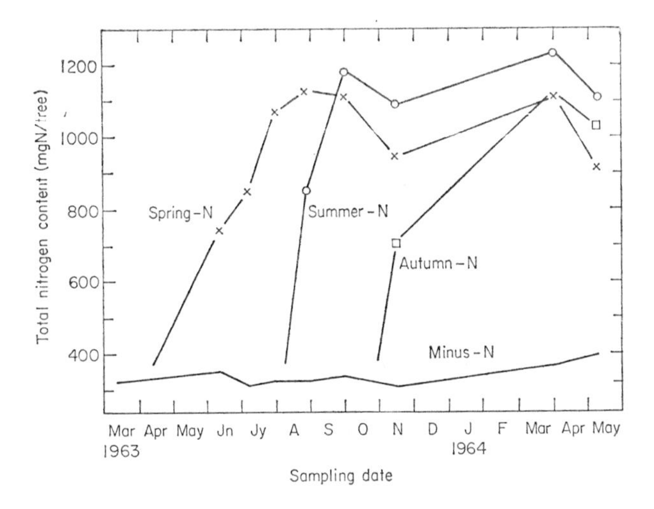

FIG. 2. The total nitrogen content of whole trees.

been studied by many workers and the general conclusion is that with conventional, i.e. late-winter, fertilizer programmes the leaves lose about half their pre-senescent nitrogen content to the perennial parts of the tree in the 3–4 weeks prior to abscission (Oland, 1963). In this experiment, for both the "minus-N" and "spring-N" trees the scion increased in nitrogen by about 30% of the peak leaf content from the end of August to mid-November, but on the "summer-N" trees less than 5% of the maximum total leaf nitrogen returned to the scion over this period. It would appear that the renewed translocation of nitrogen to the leaves following "summer-N" applications and the consequent boost to photosynthesis there, enabled the leaves to function normally until another cause, possibly low temperature, induced abscission very rapidly, with little or no time for nutrient migration.

THE NITROGENOUS CONSTITUENTS OF FRUIT PLANTS

That amino acids are synthesized in the roots of fruit plants has been known for many years and it is now recognized that in many woody plants much, if not all, of the absorbed nitrogen is incorporated into amino acids, amides or ureides in the roots and is translocated as such to the rest of the plant (Bollard, 1957a, b and c).

The most efficient form of nitrogen to use as fertilizer has been the subject of prolonged research; and for apple trees in soil the choice appears to be dependent on rainfall, as nitrate is more readily leached from soil;

on soil pH, as the ammonium ion is a poor source of nitrogen in acid soils; and on the cover crop, due to competition. (Magness *et al.*, 1948; Sahulka, 1962; Van der Boon *et al.*, 1962.)

It is generally accepted that apple trees in the field use principally nitrate ions, and that ammonium ions are normally oxidized biologically in the soil beforehand, hence the importance of soil pH referred to above. Of particular interest, therefore, are the results of experiments with an inert medium and sterile nutrient solutions which showed that trees given nitrogen only as the ammonium ion made no more growth than those given none, although analysis showed that the former had in fact absorbed a considerable amount of nitrogen (Grasmanis and Leeper, 1965). Further work has suggested that the presence of ammonium ions in the tree roots inhibits nitrate reductase by a feed-back mechanism (Grasmanis and Nicholas, 1966).

Under field conditions, only organic forms of nitrogen have been found within apple trees away from the fine roots, nevertheless recent reports claim that significant amounts of nitrate also were detected in these regions from trees grown in nutrient culture solutions (Özerol, 1965; Titus and Özerol, 1966). Other fruit plants that have been examined include the pear (Lewis *et al.*, 1963), the peach (Schneider, 1958; Taylor, 1967) and the almond (Kazaryan and Karapetyan, 1962), all of which contained predominantly organic nitrogen, while the grape-vine has been found to translocate both nitrate and amides (Cook and Kishaba, 1956; Isoda, 1962).

Freeze-dried samples of all the stock-wood material from the experiment on the time of application of nitrogen, described in previous sections, have been extracted and the main soluble constituents determined (D. G. Hill-Cottingham and Denise Britton, unpublished). Extractions were carried out with 0·02 N HCl in 70% aqueous ethanol for 24 hours at −15°C. The results (Table II) show that in the "minus-N" trees there was some hydrolysis of insoluble material before budbreak, but thereafter during the growth of the scion the soluble nitrogen fell to a very low level while that of the insoluble fraction fell less markedly. Oland (1959) proposed that the soluble nitrogen functions mainly as reserve to be used for growth and that a part of the insoluble nitrogen can also serve as such. In contrast, after the addition of fertilizer nitrogen, the levels of both soluble and insoluble fractions were increased rapidly even during scion growth, the response in the soluble nitrogen being the more rapid. The figures quoted for total amide content were obtained following hydrolysis, but chromatography has shown that this was predominantly asparagine. Both Oland (1959) and Sahulka (1962) had also found asparagine and arginine to account for the greater part of the total soluble nitrogen content in apple root or shoot material, although Romanovskaya (1963) found

TABLE II

Nitrogenous constituents of the stock wood. Results expressed as the percentage N found as amide or as arginine, etc.

Treatment	Minus-N				Spring-N				Summer-N				Autumn-N			
Sampling date	Soluble			Insol.	Soluble			Insol.	Soluble			Insol.	Soluble			Insol.
	Amide	Arg.	Other		Amide	Arg.	Other		Amide	Arg.	Other		Amide	Arg.	Other	
1963 19–25 Mar.		0·099		0·220												
10–12 June	0·030	0·082	0·025	0·173	0·162	0·170	0·028	0·240								
8–10 July	0·013	0·018	0·011	0·148	0·190	0·264	0·037	0·374								
29–30 July	0·009	0·014	0·011	0·136	0·145	0·352	0·022	0·376								
26–29 Aug.	0·005	0·004	0·010	0·121	0·115	0·344	0·020	0·361	0·123	0·096	0·014	0·222				
30 Sept.–2 Oct.	0·007	0·006	0·010	0·142	0·087	0·332	0·018	0·358	0·091	0·197	0·015	0·284				
18–21 Nov.	0·009	0·021	0·012	0·163	0·052	0·261	0·009	0·378	0·051	0·311	0·015	0·352	0·094	0·104	0·010	0·231
1964 31 March–8 April	0·011	0·045	0·008	0·156	0·031	0·345	0·016	0·408	0·044	0·324	0·017	0·301	0·106	0·193	0·012	0·234
4–8 May	No freeze-dried sample				0·048	0·263	0·010	0·302	0·045	0·394	0·023	0·364	0·118	0·205	0·021	0·283

some additional major constituents at times during the annual cycle of the current season's growth.

Sahulka (1962) has assigned different roles to asparagine and arginine, the former as a mobile form of nitrogen and intermediate for protein synthesis, the latter more as a reserve. The results presented in Table II for the "spring-N" and "summer-N" material are in keeping with this hypothesis, for they show that the asparagine level falls and that of arginine rises as the tree approaches the dormant season.

The same author also concluded that flower-bud initiation does not depend on the content and ratio of nitrogenous compounds in the trees but more on the mechanisms controlling growth and on nutrition. The results of the present investigation support the view that there is no one nitrogenous compound, present in concentrations detectable by the methods used, that is responsible for flower initiation, for in fact there were no significant differences between treatments in the percentage of flower buds initiated. But the possibility remains that the concentration of certain amino acids present at particular stages during development could affect the quality of the resultant flower.

The percentage of soluble nitrogen present in the leaves is always less than that in other parts of the tree. Also, in apple leaves asparagine and arginine are present only in relatively small amounts, the bulk of the soluble nitrogen being made up of aspartic acid, glutamic acid, glutamine, alanine or serine at different times during the season (McKee and Urbach, 1953; Bollard, 1957a; Sahulka and Silova, 1960).

Most of the amino acids commonly present in proteins have been found in at least small or trace amounts in all fruit tree material. Certain tissues however, apparently, at times accumulate relatively large amounts of the less-common amino acids, for example, proline in the pollen (Tupy, 1963) and tryptophan in the seeds (Chvojka *et al.*, 1962). Some non-protein amino acids and amines have also been found, notably in fruit buds or spurs (Bielińska-Czarnecka, 1963; Briner and Grasmanis, 1964).

The method of vacuum extraction of tracheal sap from apple shoots has been used to study the seasonal fluctuations in the nitrogenous constituents from a number of rootstocks and varieties (Bollard, 1957a) and also the effects on sap compositions of ground application of fertilizer nitrogen in either spring or summer (Hill-Cottingham and Bollard, 1965). In both investigations aspartic acid, asparagine and glutamine were found to be the major sap constituents at all times, with arginine and glutamic acid also becoming important at the beginning or end of the season. One unexpected result was that, following the spring fertilizer application, the percentage composition of the sap from both treated and untreated trees remained approximately the same, in spite of a doubling of concentration in the former. It has been suggested that in the apple all assimilated nitrogen passes through a transitory storage phase, that the nitrogenous compounds in the sap are derived from reserve material and that addition of the fertilizer merely accelerates this mobilization. In the description of his technique, Bollard (1953) stated his belief that the sap thus obtained originally contained most of the nutrients coming from the roots although some losses to other tissues would have occurred before it reached the young shoots from which it was extracted. Recent developments emphasize the reality and magnitude not only of the losses due to the selective absorption of certain amino acids from the transpiration stream by the lower leaves of a plant stem (Pate *et al.*, 1964), but also of the gains as these compounds are later re-exported up the stem, possibly after metabolic changes (Pate *et al.*, 1965). Hence if the changes in a tree shoot are similar to those in an annual plant, it would appear that the tracheal sap as extracted is not necessarily representative of the sap leaving the roots.

PRACTICAL CONSIDERATIONS AND CONCLUSIONS

The results of the pot experiment described above would appear to support the use of nitrogen fertilizers in summer or autumn, but it would be misleading to translate them from the much-simplified plant system that was used in this study to normal cropping trees in England or in other climates.

Under severe conditions trees given late-season nitrogen have been reputed to be more susceptible to winter injury (Edgerton, 1957), especially where vegetative growth has been prolonged. Romanovskaya (1963) found that the degree of winter hardiness was inversely correlated with the asparagine concentration in the apple shoots during the period of deep dormancy. It is thus of interest that the results in Table II show that the asparagine level did in fact remain higher through the winter following "autumn-N" than following the other treatments.

The most important factor that was not included in the experiment was the effect of the presence of an existing crop. The existence of active growing points in the seeds of the developing fruits would almost certainly affect those in the flower buds developing for the following season. This effect could well be reciprocal and take place by straight-forward competition for any additional nutrients and possibly by growth substance interactions.

The most widespread objections to the use of fertilizer nitrogen during the summer are due to its unfavourable effect on fruit colour and storage quality. These disadvantages could not result from fertilizer applications at or after fruit picking and these aspects need re-examination.

It is therefore concluded that, particularly where tree nitrogen is a limiting factor, fertilizer nitrogen applied in the summer or autumn in this country can be advantageous for flower quality and crop yield. The effects of these fertilizer applications on the existing crop and on future crops can best be determined by further experiments combining the several disciplines concerned.

REFERENCES

Bielińska-Czarnecka, M. (1963) *J. Sci. Fd Agric.* **14**, 527.
Bollard, E. G. (1953). *J. exp. Bot.* **4**, 363.
Bollard, E. G. (1957a). *Aust. J. biol. Sci.* **10**, 279.
Bollard, E. G. (1957b). *Aust. J. biol. Sci.* **10**, 288.
Bollard, E. G. (1957c). *Aust. J. biol. Sci.* **10**, 292.
Boynton, D. and Oberly, G. H. (1966). *In* "Nutrition of fruit crops". (N. F. Childers, ed.) p. 1. Horticultural Publications, New Brunswick.
Briner, G. P. and Grasmanis, V. O. (1964). *Nature, Lond.* **202**, 359.
Chvojka, L., Veres, K. and Kozel, J. (1961). *Biol. Pl. Prague* **3**, 140.
Chvojka, L., Travnicek, M., Chaloupka, J. and Rihova, L. (1962). *Biol. Pl. Prague* **4**, 315.
Cook, J. A. and Kishaba, T. (1956). *Proc. Am. Soc. hort. Sci.* **68**. 131.
Delap, Anne V. (1967). *J. hort. Sci.* **42**, 149.
Edgerton, L. J. (1957). *Proc. Am. Soc. hort. Sci.* **70**, 40.
Grasmanis, V. O. and Leeper, G. W. (1965). *Agrochimica* **10**, 54.
Grasmanis, V. O. and Nicholas, D. J. D. (1966). *Pl. Soil* **15**, 461.
Hill-Cottingham, D. G. (1963). *J. hort. Sci.* **38**, 242.
Hill-Cottingham, D. G. and Bollard, E. G. (1965). *N.Z. Jl. agric. Res.* **8**, 778.
Hill-Cottingham, D. G. and Williams, R. R. (1967). *J. hort. Sci.* **42**, 819.
Isoda, R. (1962). *J. hort. Ass. Japan* **31**, 123.
Kazaryan, V. O. and Karapetyan, K. A. (1962). *Biol. Pl. Prague* **4**, 269.
Leopold, A. C. (1964). *In* "The Hormones" (G. Pincus, K. V. Thimann and E. B. Astwood, eds.) Vol. 4, p. 1. Academic Press, London and New York.
Lewis, L. N., Tolbert, N. E. and Kenworthy, A. L. (1963) *Proc. Am. Soc. hort. Sci.* **83**, 185.
Loeffler, J. E. and van Overbeek, J. (1964). *In* "Regulateurs naturels de la croissance végétale. (J. P. Nitsch, ed.) Colloques internat. du C.N.R.S., Paris. No. 123, 77.
McKee, H. S. and Urbach, G. E. (1953). *Aust. J. biol. Sci.* **6**, 369.
Magness, J. R., Batjer, L. P. and Regeimbal, L. O. (1948). *J. agric. Res.* **76**, 1.
Mason, A. C. and Whitfield, A. B. (1960). *J. hort. Sci.* **35**, 34.
Mori, H. and Yamazaki, T. (1959). *Bull. Tohoku natn. agric. exp. Stn* **15**, 69.
Oland, K. (1959). *Physiologia Pl.* **12**, 594.
Oland, K. (1963). *Physiologia Pl.* **16**, 682.
Özerol, N. H. (1965). *Diss. Abstr.* **25**, 4340.
Pate, J. S., Wallace, W. and Die, J. van, (1964). *Nature, Lond.* **204**, 1073.
Pate, J. S., Walker, J. and Wallace, W. (1965). *Ann. Bot.* **29**, 475.
Poulsen, E. and Jensen, J. O. (1964). *Tidsskr. PlAvl.* **68**, 477.
Romanovskaya, O. I. (1963). *Soviet Pl. Physiology*, **10**, 581.
Sahulka, J. (1962). *Biol. Pl. Prague* **4**, 3.
Sahulka, J. (1962). *Rostlinna Vyroba* **8**, 617.

Sahulka. J. and Silova, A. (1960). *Biol. Pl. Prague* **2**, 70.
Schneider, A. (1958). *C.r. hebd. Séanc. Acad. Sci. Paris* **247**, 1034.
Taylor, B. K. (1967). *Aust. J. biol. Sci.* **20**, 379.
Titus, J. S. and Özerol, N. H. (1966). *Proc. XVII Int. hort. Congr. Maryland*, Vol. I. Summary of Paper No. 161.
Tupy, J. (1963). *Biol. Pl. Prague* **5**, 154.
Van der Boon, J. A., Pouwer, A. and DeVos, N. M. (1962). *Proc. XVI Int. hort. Congr. Brussels*. Vol. III, 151.
Williams, R. R. (1963). *J. hort. Sci.* **38** 52.
Williams. R. R. (1965). *J. hort. Sci.* **40**, 31.

Discussion on Section 3

Regarding Dr L. C. Luckwill's reference to chlorate toxicity being due to its reduction to chlorite, Dr Sims commented: "We are interested in chlorate as a compound with a structure similar to nitrate in an attempt to find in yeast a permease. Dr. A. H. Bussey grew a yeast for a long time in the presence of chlorate and nitrate and in no way did it affect the growth. He isolated an enzyme and showed that chlorate is enzymically reduced to chlorite with NADP, and chlorite was reduced in the same way with the nitrite reductase system. If the level of nitrate and nitrite reductase systems in yeast is measured, there is a 70–80 % excess activity over requirement; the measure of control is a restraint being applied to the enzyme. The result of growing in chlorate would merely lead to a reduction in the pool of compounds that give a feed-back restraint on the enzyme. One would not, under the circumstances, expect very much effect on the overall nitrogen metabolism. The position in plants may be very different as to whether or not there is any feed-back control on the enzyme."

Dr Folkes commented on the effects of growth-controlling compounds such as 2,4-D on the induction of nitrate reductase: "I would predict that the second of Beevers' suggestions—that involving nucleotide metabolism—is the correct one. We have found that nucleotide availability is probably an important limiting factor in the control of protein synthesis and growth; yeast in a turbidostat shows large increases in growth rate when changes in conditions, such as in a temperature transition, result in a temporary increase in the nucleotide pool. We know also from our own work that amino acids are the repressors of nitrate reductase, so it seems likely that the effects of 2,4-D might be explained in terms of an alteration in the balance of nucleotide and amino acid synthesis leading to increased protein synthesis and a consequent decrease in the amino acid pool. Under these conditions the repression of nitrate reductase would be less severe."

Dr Luckwill agreed that there was much evidence showing increase of protein concentration in plants following 2,4-D treatment.

Following Dr J. S. Pate's paper, the Chairman asked: "Do you always find relatively low levels of aspartate in leaves, particularly in relation to the assimilation of labelled CO_2? In most of the leaves I have been looking at recently, including barley, there are fairly high levels of aspartate and very rapid formation from $^{14}CO_2$ in the light."

Dr Pate replied: "Aspartic acid is an uncommon constituent of the leaves of field pea, except in cases of nitrogen deficiency when it is present in large amounts, presumably since reserves of asparagine are being utilized as a source of nitrogen. It is interesting that in the field pea, roots are relatively rich in aspartic acid but low in glutamic acid whereas in the leaves the reverse is true. Other species, in our experience, may be different, showing very high levels of aspartate in leaves. As far as the formation of amino acids by photosynthesis is concerned, aspartic acid and aspartyl residues of protein are poorly labelled in photosynthesis (Fig. 2, p. 222). In the field pea, large quantities of asparagine and aspartic acid arrive in the leaves from the roots, and the leaf's requirement of aspartate and related compounds may be met in this way."

Dr O. R. Jewiss said that Dr Pate had indicated very clearly how the stem apex receives a uniform supply of amino acids derived from the photosynthetic products of the leaves. Had he looked at the effect of removing a leaf in order to see how the supply of amino acids to the stem apex is affected and how quickly the contribution from remaining leaves readjusts to compensate for the leaf removed?

Dr Pate answered: "We have not done this. However, if a plant of field pea is decapitated leaving one or two lower leaves attached to the stem stump these leaves immediately accumulate large amounts of free amino acids and synthesize more chlorophyll and protein. Also, within 24 hours there will be a return in great strength of nitrate reductase, an enzyme which had declined at an early stage. This type of compensatory mechanism might operate if a single leaf were removed from a shoot."

In answer to Dr T. A. Smith's request for information regarding the origin of the metabolites utilized by the developing seed and the contribution made by the pod to these, Dr Pate commented: "The pod does contribute certain organic compounds to the seed, particularly nitrogenous compounds. However, much of the carbohydrate requirement of the seed is met by currently photosynthesized sucrose, particularly that formed in the leaf which subtends the pod."

Dr D. O. Gray then asked: "Where is tryptophan synthesized in the plant since this compound appears in the amino-acid-analyzer traces?"

Dr Pate said that no clear idea was formulated although trace amounts of free tryptophan had been recovered from all parts of the plant.

To Dr Gray's further question: "Does the total leaf protein really contain as little as 1% cysteine?" he replied: "The content of cysteine does seem to be lower than in reports by other workers. Figure 14, p. 236, records it as approximately 2·5% of the total carbon in leaf protein."

Arising from the contribution of Dr D. G. Hill-Cottingham, Dr Hewitt said: "One of the striking differences in the effect of nitrogen supply was on the actual fruit crop produced. You had a very big blossom formation with summer nitrogen, but not very good fruit production, whereas with autumn nitrogen, fruit production was very high. Is this connected with arginine level? With adequate nitrogen, arginine is high where spring and summer nitrogen had been applied but is very much lower, although still rising, with autumn nitrogen. If we can leave out of consideration the minus-nitrogen series, which is something we can do for this question, is arginine involved in fruit set as opposed to blossom formation?"

Dr Hill-Cottingham replied: "The short answer is that we know practically nothing about the requirements of the individual stages of development of the apple bud, but I do not believe that the level of arginine, or of any major nitrogenous constituent, is alone the critical factor for fruit set. Nutrition and growth substances are so inter-related that one should not be considered without the other. We would attribute the high fruit set on the "autumn-N" flowers primarily to the long period over which their ovules remained viable during blossoming. This in turn resulted from the stimulation of these ovules by a root-produced substance, probably supported by favourable nutrition before or during blossoming. The corresponding response by the roots to "spring-N" or "summer-N" would have reached the buds on those trees soon after fertilizer application and hence that stimulus would have been used largely in the development of earlier stages (Fig. 1, p. 220)."

Conclusion

E. W. YEMM

As our Chairman, Lord Waldegrave, pointed out, we have had a varied symposium covering many aspects of the assimilation and metabolism of nitrogen in plants. The topics have ranged from the basic chemistry of nitrogen to biochemical and physiological mechanisms and their significance in the growth and development of plants. At this Research Station, which has made so many valuable contributions to the study of plant nutrition, there is little need for me to emphasize the key position of nitrogen as a plant nutrient and the importance to agriculture and horticulture of a better understanding of its assimilation and metabolism.

At this stage of the proceedings, I shall not attempt to give a comprehensive summary of the papers which have been presented: such an attempt would obviously involve a rather superficial review of some fourteen papers which have already been discussed in detail and are available as pre-prints. As "tail-end Charley", I can perhaps claim some room for manoeuvre; I propose to discuss briefly one or two problems of nitrogen metabolism which seem to me of particular interest in relation to the general physiology of plants and plant organs. In several of the papers, reference has been made to biochemical mechanisms whereby the assimilation of nitrate and the biosynthesis of amino acids and protein may be coupled to respiratory and photosynthetic processes. It is upon this rather wide subject that I should like to make a few concluding comments.

Until comparatively recently, it has generally been assumed that the assimilation of nitrogen and synthesis of amino acids were only indirectly dependent upon photosynthetic activity. Fairly stable carbohydrates, such as sucrose and starch, were regarded as the chief products of photosynthesis, their subsequent breakdown in respiratory metabolism constituting a powerful link with the endergonic reactions involved in the metabolism of nitrogen and other biosynthetic activities. Indeed, there is much evidence that this relationship dominates the assimilation of nitrogen in roots and other non-photosynthetic plant organs. Some years ago my colleague, Dr Willis, was able to show in experiments with excised roots of barley that the assimilation of ammonium salts and nitrates was closely coupled with the breakdown of carbohydrates in respiration. The use of isotopic nitrogen (^{15}N) clearly indicated that the main pathway of assimilation led to the synthesis of glutamic acid and glutamine, and that the lag in nitrate assimilation involved adaptive enzyme formation in young roots. More recently, Dr Berner, working with us, has found that the level of amino acids in young seedlings may have a controlling influence on their assimilation of external sources of nitrogen. Many close parallels can in fact be drawn between the phenomena of nitrogen assimilation in roots and those in heterotrophic microorganisms, such as food yeast, fully discussed by Drs Folkes and Sims during the symposium. It may well be that similar regulating mechanisms control the activity of glutamic acid dehydrogenase and other enzymes operating in the early stages of assimilation. Dr Miflin's paper has clearly pointed to the mitochondria of root cells as an important centre of these activities and has again emphasized the close relation between cell respiration and the assimilation of nitrogen.

Against this background, Dr Pate has given us a valuable account of the circulation of nitrogenous and other metabolites in growing plants of the field pea. In this species an extensive translocation of sucrose from the mature leaves to the roots sustains the assimilation of nitrogen and the biosynthesis of amino acids, particularly glutamic and aspartic acids and their amides. Under normal conditions the development of the shoot system depends, at least in part, upon the supply of amino acids and amides reaching it from the roots. At the same time the shoot system, especially the young leaves, may make an important contribution to the synthesis and interconversion of amino acids. In this regard the formation of alanine, glycine and serine in leaves during photosynthesis appears to be complementary to the activities of roots. Our own experiments with barley leaves, supplied with labelled carbon dioxide ($^{14}CO_2$), have shown that aspartic acid may also be rapidly formed during photosynthesis. There is little doubt that young expanding leaves are one of the most active centres of nitrogen metabolism in growing plants; the possibility that the synthesis of amino acids and proteins, much of it occurring in the plastids, may be closely coupled to photosynthesis by means of photoreduction and photophosphorylation has been briefly considered elsewhere. It seems probable that in young leaves the rapid synthesis of distinctive proteins and other structural components of chloroplasts may represent a phase in their development which precedes that in which sugars and starch are the chief stable products of carbon dioxide assimilation.

Evidence that photosynthesis plays an important part in the assimilation and utilization of nitrogen comes from several other directions. The reduction of nitrate and nitrite in leaves is light-dependent. Some of the underlying biochemical mechanisms for nitrite were discussed in detail by Dr Hewitt and his collaborators. Their purification of nitrite reductase and studies of its action strongly suggest that under physiological conditions ferredoxin acts as a highly efficient electron donor in a photochemical system with chloroplast grana. The supply of carbon skeletons for synthesis of amino acids in leaves may also depend directly on photosynthesis. Extensive data now available on the assimilation of labelled carbon dioxide by leaves and autotrophic algae indicate that under favourable conditions a major part of the carbon dioxide is rapidly utilized in the synthesis of amino acids. Some years ago Bassham and others suggested, from an analysis of such data, that a relatively independent metabolic pool of amino acids is formed in the chloroplasts of algae and that distinctive mechanisms of amino acid synthesis may operate here by the diversion of labile intermediates of the photosynthetic cycle.

It is, as yet, no means certain to what extent photosynthetic and respiratory metabolism are independently organized in the leaf cells of higher plants. The early measurements of total gaseous exchanges of leaves in the light and in the dark, made by means of isotopic oxygen (^{18}O), indicated that respiration continued in the light at about the same rate as in the dark. But there is now a good deal of evidence that the respiratory mechanisms may be considerably modified in the light. For example, it has been shown recently in several laboratories that if leaves are allowed to assimilate ^{14}C-labelled carbon dioxide for a preliminary period and the release of carbon dioxide is then measured in the light and in the dark, there are substantial differences in the labelling of the carbon dioxide produced. In the light, its specific activity approaches that of the carbon dioxide recently assimilated by the leaves, whereas in the dark the specific activity is much lower. This, and other evidence, suggests that photosynthetic intermediates formed in chloroplasts in the light may be rapidly released and drawn into respiratory metabolism. The term "photorespiration" has sometimes been used to denote the distinctive features of respiration in the light. Relatively little is known of the underlying biochemical mechanisms, but the most plausible hypothesis that can be offered at present is that glycollate and glyoxylate may be important reactants linking photosynthesis and respiration in leaves. It seems likely that both of these metabolic systems contribute to the assimilation of nitrogen and synthesis of amino acids in photosynthetic plant organs.

It will be evident from these few general comments that much remains to be discovered in the nitrogen metabolism of plants, both at the physiological level of the growing plant and its organs, and at the cellular and biochemical level. The versatile biosynthetic processes of autotrophic plants present, in many ways, unique problems, particularly in relation to photosynthetic activities; it is here that the guidance of comparative biochemistry must be used with caution. As Professor Davies pointed out in his discussion of amino acid metabolism, advances in some directions have been disappointingly slow, but there is no doubt that symposia such as this give an important stimulus to progress.

Members of the symposium would wish to express their thanks to the Director, Professor Kearns, for the excellent way in which he has helped.

Much of the value of symposia depends on opportunities for informal discussion, which has been an outstanding feature here and which reflects the very effective way in which Dr Hewitt has organized the topics and speakers. Something of the efficiency of the organization can be seen from the almost complete list of pre-prints which have been available, and for this and other detailed arrangements we are indebted to many members of the staff, particularly Dr Cutting and Mr Clothier. For the admirable domestic arrangements we have enjoyed, our appreciation is expressed to Mr A. B. Reynolds, the Warden of Churchill Hall, and to Mrs M. J. Swatridge at Long Ashton.

Author Index

Numbers in *italics* indicate pages which contain reference lists

Subject Index

This index is a guide to main arguments and topics of special interest

D

E